Plant Mites and Sociality

Yutaka Saito

Plant Mites and Sociality

Diversity and Evolution

Springer

Dr. Yutaka Saito
Professor of Animal Ecology
Hokkaido University
Research Institute of Agriculture
Kita 9, Nishi 9, Kita-ku
Sapporo, Hokkaido 060-8589
Japan

ISBN 978-4-431-54746-4 ISBN 978-4-431-99456-5 (eBook)
DOI 10.1007/978-4-431-99456-5
Springer Tokyo Berlin Heidelberg New York

Front cover:
Top: Male prey counterattacking larval predator
Lower left: Two-spotted spider mite, *Tetranychus urticae*

Back cover:
Upper left: Social cheyletid mite waiting for prey in a sit-and-wait stance
Upper right: Symbiosis between a spider mite and a psyllid?

Background, front and back covers: Sasa ground cover in Hokkaido forest

Printed on acid-free paper

Springer is part of Springer Science+Business Media (www.springer.com)

Preface

This book is on mite sociality. I have studied spider mite diversity for 35 years mostly from the perspective of evolutionary ecology, and this led to the discovery of several new phenomena on sociality in plant mites. My main interest is similar to that of Jean-Henri Fabre, the famous French entomologist, although my studies have been restricted to a few mite groups and yield no comparison to his broad, encyclopedic knowledge. On the other hand, I used some modern methodologies that could not have been adopted in his period. Of course I would not dare to compare my contributions to those of such a great entomologist, but I believe that the studies described in this book bear some resemblance to his work. At least I have been fortunate to learn about the wonderful world of mites, and I hope the readers of this book will experience this excitement as well.

When I first planned to write this book, several colleagues advised me not to focus just on my own work but to address wider themes in mite biology. I have surely tried to do so, but now in retrospect I am wondering whether I achieved this goal, because my target mite species are limited to a narrow area of Asia and may not be representative of all mites. However, I suspect that these "extraordinary" species may well be distributed over a much larger part of the world: they may simply not have been discovered and studied so far. Therefore, if someone wishes to study them after reading this book, it will give me considerable satisfaction.

Mites are small animals that live in a world they cannot see. They are also hard to see by the naked eye because of their very small body size. Perhaps this is why they were named "mite" instead a more impressive name like "dino-" or "dyna-" mite. However, as I will explain below, this animal group comprises a large number of species showing great diversity. My studies have been restricted to only a few groups, in particular the family Tetranychidae, which are generally known as spider mites. If I succeed in showing how diverse such a small group of mites in nature can be, it may leave the reader with an impression of how little is known and how mysterious the world of mites really is. In this book, I will address several important findings in spider mite evolution, species radiation, and behavioral ecology. These findings may reflect the complex evolutionary history of these taxa and also help to point out clearly what is known and what is not yet known to date. I believe that such a book could be useful for young researchers with an interest in this group. In order to advance understanding, I have created "boxes" to explain issues raised briefly in the text, and at the end of the book I have provided appendices to explain the details of particular topics and/or terms. I hope readers will find these features of the book useful as background material.

My studies have been supported by many co-researchers: Drs. K. Sahara, T. Takada, K. Takahashi, Mh. Osakabe, S. Kudo, H. Nakao, Y. Yasui, K. Mori, T. Sakagami, A.R. Chittenden, Y. Sato, K. Ito, N. Tsuji, Y.-X.Zhang, J.-Z. Lin, M. Kongchuensin, as well as Mr. J. Ji, Mr. Yanagida, Ms. M. Nishijima, Mr. M. Horita, Mr. T. Yoshida, Mr. J. Yano, Mr. Y. Nakano, Mr. T. Ogawa, Ms. A. Kikuchi-Ise, and Ms. M. Kanazawa. The late Prof. Emeritus H. Mori was my teacher who introduced me to the study of mites. The late Prof. Emeritus S. Sakagami, the late Prof. Emeritus S. Ehara, Dr. U. Gerson, Dr. Y. Ito, Dr. H. Abe, Dr. S. Takagi, Dr. T. Kumata, Dr. M. Sasagawa, and Mr. S. Moriyama also were my teachers who supported my studies. Dr. S. Akimoto, Dr. T. Endo, Dr. N. Ohsaki, Dr. T. Gotoh, Dr. J. Takabayashi, Dr. N. Yamamura, Dr. A. Yamauchi, Dr. J.C. Choe, Dr. W. Lee and Dr. Y. Watanuki are my colleagues and/or friends who provided much valuable advise. Dr. M.W. Sabelis kindly reviewed and revised this book manuscript thoroughly. My wife, Keiko Saito, has stimulated my career in science for such a long time. I wholeheartedly thank all of them for encouraging me to write this book.

I very much appreciate Dr. K. Kikuzawa's recommending me to the publisher, Springer Japan, and I am grateful to Ms. A. Hiraguchi and Mr. K. Hashimoto who made a great effort to publish this book. Financial support was received from Grants-in-Aid from JSPS (nos. 11640626, 13440227, 13575021, 17370005, 18657007, 20370006), MEXT, through Special Coordination Funds for the Promotion of Science and Technology entitled "Hokkaido University Sustainability Governance Project" and from Nissei Zaidan, a private Japanese fund.

Yutaka Saito
Sapporo, Japan

Contents

Chapter 1
What Are Mites?

The purpose of this book is not to provide a handbook of mites or a monograph of the subclass Acari, because I would not be capable of this nor do I wish to do so. If readers intend to learn about Acari in general, I recommend reading two important and comprehensive books written by Krantz and Walter (2009) and Walter and Proctor (1999). I recommend consulting Fig. 1.1 when you have any doubts on the basic classification of mites.

1.1 Brief Systematic Overview of Acari

The Acari represent a subclass of the class Arachnida. They are divided into two orders: Parasitiformes and Acariformes. Although there have been several discrepancies in the classification of higher taxa, I consistently adopted the classification of Krantz and Walter (2009) in this book, this being the most standard treatise on systematics of higher taxa above family level. In this book, I mainly address the mite species of two suborders: the Actinedida (Prostigmata), to which belong the families Tetranychidae, Eriophyidae, Cheyletidae, and Tenuipalpoidini; and the Gamasida (Mesostigmata), to which belong the families Phyotoseiidae, Laelapidae, and Varroidae. I have omitted the authority names in the text when describing species to improve the readability of the text, hoping that taxonomists will not criticize me for this approach. Instead, I have attempted to spell out the full species epithets in the tables whenever possible.

1.2 Peculiarities of Mites for Ecological and Evolutionary Studies

The features that discriminate Acari (mites and ticks) from other arthropods derive from their special morphological characters. Among these, body size should receive attention first, because most mites are smaller than 1 mm (Walter and Procter 1999),

Y. Saito, *Plant Mites and Sociality: Diversity and Evolution*
DOI 10.1007/978-4-431-99456-5_1, © Springer 2010

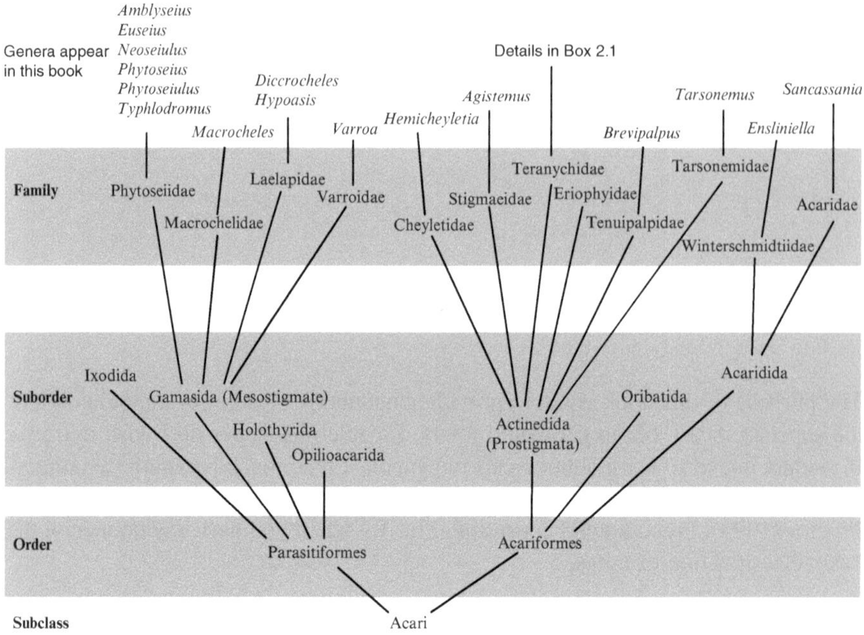

Fig. 1.1 Brief classification of subclass Acari according to Krantz et al. (2009). Only main groups involving species addressed in this book are shown

ranging from 0.13 mm in Eriophyidae and Tarsonemidae to 2.4 mm in Trombidiidae (ticks are bigger than mites, and some are more than 6 mm in body length). As mentioned, the Acari are divided into two large groups, i.e., by general names, ticks (suborder Metastigmata) and mites (suborders Mesostigmata, Prostigmata, Astigmata, and Cryptostigmata). In this book, I only consider the latter group. Mites (as well as ticks) have bodies without apparent distinction of head, thorax, and abdomen (Fig. 1.2). Furthermore, they have four pairs of legs and no wings. These basic peculiarities make mites sometimes easy or sometimes hard to study, depending on the topic of interest. To show what are the unique and useful points of mites and what are the difficulties in their study is one of the themes in this book.

Mites having small bodies (Fig. 1.2), less than 1 mm in length, are affected by gravity, but even more by atmospheric buoyancy once they become airborne. This may be a reason why they have not evolved wings, because it is very difficult to develop organs to deal with air resistance (Schmidt-Nielsen 1984). The aggregative life style of mites is thus thought to be a consequence of their relatively low mobility. Furthermore, most mites have no functional eyes, although several taxa have eye-spots (many Prostigmata species). Even if species have eye-spots, they only sense photo-intensity and light direction (McEnroe 1969; McEnroe and Dronka 1969). Thus, mites basically live in a world that they cannot see but only explore by tactile and olfactory means (Sabelis and Dicke 1985; Takabayashi and Dicke 1996; Dicke et al. 1998; Radwan et al. 2002; Sato et al. 2003).

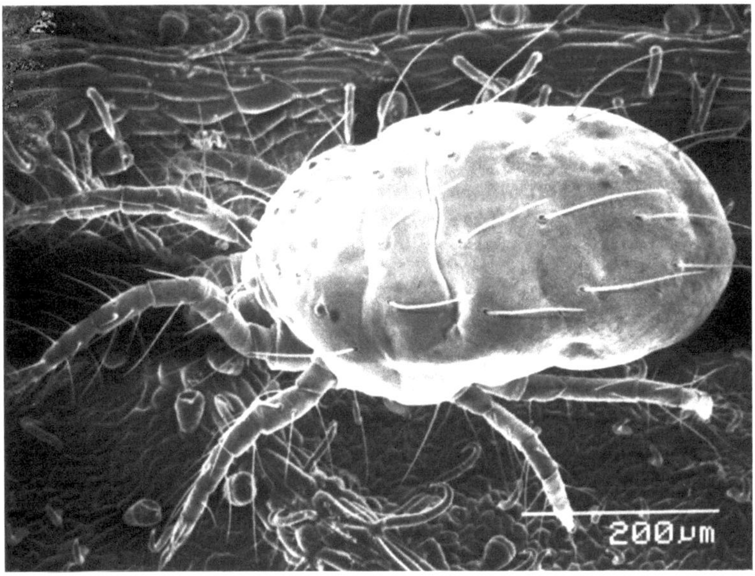

Fig. 1.2 Two-spotted spider mite, *Tetranychus urticae* Koch, on kidney bean leaf, the most famous species worldwide as an agricultural pest. {Micrograph by Y. Saito under scanning electron microscope (SEM) with cryo-system attached}

Perhaps because they are wingless, some mite species have silken threads to assist in movement and attachment (Saito 1977a; Walter and Proctor 1999). This ability must diversify the lives of several mite taxa in their habitats, as described hereafter. Spider mites, the main "cast" of this book, typically produce silken threads and use these in various ways that represent an adaptation to survive on the leaf surfaces of their host plants.

Mitchell's (1973) study on the life history of *Tetranychus urticae* in connection with the *r*/*K* selection theory, Helle and Overmeer's (1973) work on genetics, and the work by Potter et al. (1976) on sexual selection are considered to be classic studies on spider mite evolution. After that, various studies on spider mites were reviewed in the book *Spider Mites. Their Biology, Natural Enemies, and Control* (Helle and Sabelis 1985), published about 24 years ago. In the same book, Crozier (1985), a researcher famous for his work on social insects, stressed the availability of spider mites as model organisms for studying evolutionary ecology and genetics. I also had the honor of participating in this book, addressing the diversity of spider mite life types in relation to the use of silk for constructing webs and the associated defecation behavior (Saito 1985). At that time, diversity was studied quite separately from these behavioral features, but I now believe that these two are really closely associated with each other.

Chapter 2
Plant Mites

"Plant mites" are not a taxonomically defined group, but a collective term that is usually denoted as "plant-inhabiting mites." This group consists of many mite species belonging to different taxa, such as the phytophagous mite families Tetranychidae, Eriophyoidea (superfamily) and Tenuipalpidae, and the predacious mite families Phytoseiidae, Stigmaeidae, Cheyletidae, Cunaxidae, and Anystidae. Furthermore, several mite taxa including Tydeidae, Tarsonemidae, and arboreal Oribatidae mites are also included (Table 2.1), most of which are considered to be polyphagous. On a particular plant species, researchers tend to focus on a particular mite species and to ignore the others occurring there that are from different taxa. However, as is discussed in this book, almost all mite species must interact with each other on the plant, and such interactions must be important to understand the real world where these microarthropods are living, i.e., their ecology. This is one reason why we adopt such a collective term, plant mites. Furthermore, some plant mite species other than phytophagous species interact directly with host plant species; e.g., Walter (1996) reported that leaf domatia made by host plants greatly affect the plant mite fauna, which may be regarded as a strategy of host plants. Fournier et al. (2003) also reported that plant physical structures such as domatia, depressions, and leaf roll formation by insects must have important effects on competition and predator–prey interactions. As such, the plant leaf is a kind of micro-sphere where various mites play evolutionary games, such that we call them plant mites in the same sense as aquatic and soil mites. Among plant mites, Tetranychidae and Phytoseiidae are the most general families in agriculture and forestry, and many studies have been conducted on these groups of mites so far (Jeppson et al. 1975; Helle and Sabelis 1985; Walter and Proctor 1999). How these mite groups diversified on plants and what factors have caused their evolution on particular host plants are the main themes of this book. For this purpose, I have to address many mites with their Latin scientific names and their abbreviations (Box 2.1).

2.1 Spider Mites and Plants

Spider mites (Tetranychidae) are the main group of plant-inhabiting mites (a detailed classification appears in Box 2.2). According to Bolland et al. (1998), who provided a catalogue of spider mite species of the world, 1,189 species had been described

Y. Saito, *Plant Mites and Sociality: Diversity and Evolution*
DOI 10.1007/978-4-431-99456-5_2, © Springer 2010

Table 2.1 Main plant mite families

Herbivorous (Phytophagous)	Predacious	Omnivorous
Tetranychidae	Phytoseiidae	Tydeidae
Eriophyoidea[a]	Stigmaeidae	Tarsonemidae
Tenuipalpidae	Cheyletidae	Acaridae
Tarsonemidae	Tydeidae	Arboreal Oribatidae
Acaridae	Tarsonemidae	
	Cunaxidae	
	Anystidae	

[a]Superfamily

Box 2.1 Who's who in this book (main cast)

Abbreviated	Scientific name	Remarks
Ag. iburiensis	*Agistemus iburiensis* Ehara	Specific stigmaeid predator of *Stigmaeopsis* spider mites
Ag. summersi	*Agistemus summersi* Ehara	Stigmaeid predator inhabiting *Sasa* (*Sasa senanensis*)
Am. orientalis	*Amblyseius orientalis* Ehara	Phytoseiid predator inhabiting *Sasa*
Am. sp.	*Amblyseius* sp.	Phytoseiid predator of spider mites occurring on *Sasa*
Ap. corpuzae	*Aponychus corpuzae* Rimando	Spider mite species inhabiting wide-range of bamboo plants
D. phalaenodectes	*Dicrocheles phalaenodectes*	Parasitic mite of nocturnal moth ears (tympanic organs), so-called moth ear mite
Hm. morrii	*Hemicheyletia morii* Ehara	Cheyletid predacious mite that shows cooperative hunting
Hp. aculeifer	*Hypoaspis aculeifer* (Canestrini)	Predacious mite that mainly eats collembolans on/in soil
M. muscaedomesticae	*Macrocheles muscaedomesticae* (Scopoli)	Predacious mite of housefly eggs, living in manure
N. womersleyi	*Neoseiulus womersleyi* (Schicha)	Phytoseiid predator that is an important biological control agent of spider mites
Pl. persimilis	*Phytoseiulus persimilis* Athias Henrriot	Phytoseiid predator of spider mites, a very famous biological control agent for spider mites
Pn. citri	*Panonychus citri* (McGregor)	Citrus red mite (spider mite), injurious to citrus trees
Ps. tenuiformis	*Phytoseius tenuiformis* Ehara	Phytoseiid predator inhabiting *Sasa*
St. celarius	*Stigmaeopsis celarius* Banks	Subsocial spider mite making middle-sized woven nest on several bamboo plants in Japan
St. longus	*Stigmaeopsis longus* (Saito)	Social spider mite making large woven nest on Sasa in Japan; previous names are *Schizotetranychus longus* and *Schizotetranychus celarius* long seta form.

(continued)

Box 2.1 (continued)

St. miscanthi	*Stigmaeopsis miscanthi* (Saito)	Social spider mite making large woven nest on Chinese silvergrass; previous name is *Schizotetranychus miscanthi*
St. nanjingensis	*Stigmaeopsis nanjingensis* (Ma et Yuan)	Social spider mite making large woven nest on Moso bamboo, *Phyllostachys pubescens*
St. saharai	*Stigmaeopsis saharai* Saito et Mori	Subsocial spider mite making smallest woven nest on Sasa in Japan; previous name is *Schizotetranychus celarius* short seta form
St. takahashii	*Stigmaeopsis takahashii* Saito et Mori	Subsocial spider mite making middle-sized woven nest on Sasa in Japan. Previous name is *Schizoteiranychus celarius* medium seta form
Sz. recki	*Schizotetranychus recki* Ehara	Spider mite making compartment-type nest on Sasa
Tp. bambusae	*Typhlodromus bambusae* Ehara	Specific phytoseiid predator of *Stigmaeopsis* spider mites
Tt. urticae	*Tetranychus urticae* Koch	Two-spotted spider mite injurious to many crops
V. jacobsoni	*Varroa jacobsoni* (Oud.)	Parasitic mite injurious to honeybee broods
Y. sapporensis	*Yezonychus sapporensis* Ehara	Spider mite inhabiting on *Sasa*

by the end of December 1996. In the more than 10 years since this publication, there have been many newly described species, however.

In this family, there are two subfamilies, Bryobiinae and Tetranychinae. The latter is a very famous group because it involves many pest species of agricultural crops, orchards, and forest trees. Within the subfamily Tetranychinae, there are three tribes, Tenuipalpoidini, Eurytetranychini, and Tetranychini, and the species belonging to the latter two tribes form the main "cast" of this book. Although it is taxonomically not accurate, I refer simply to "spider mites" as a collective name for the last two tribes. Thirty-two genera are known in spider mites: 11 in Eurytetranychini and 21 in Tetranychini (Bolland et al. 1998).

Ehara (1999) reviewed the Japanese species of Tetranychidae and showed that there are three genera in the Eurytetranychini and eight (seven in accordance with Bolland et al. 1998) genera in the Tetranychini. In these two tribes, four species in the former family and 66 species in the latter family have been described (Ehara 1999). Since 1999, one genus and several new species have been added to the Tetranychini (e.g., two species of *Stigmaeopsis* were newly described by myself). The present book is not a taxonomic review. Here, it suffices to know that there are more than 1,000 species of Tetranychidae in the world and also more than 70 species in Japan.

Box 2.2 Classification of Tetranychidae (Bolland et al. 1998)

Family	Subfamily	Tribe	Genera
Tetranychidae			
	Bryobiinae	Bryobini	
		Hystrichonychini	See Bolland et al. (1998)
		Peterobini	
	Tetranychinae		
		Eurytetranychini	
			Atetranchus, Synonychus, Eurytetranychus, ***Eurytetranychoides****, Eutetranychus, Meyernychus,* ***Aponychus****, Paraponychus, Sinotetranychus, Duplanychus*
		Tenuipalpoidini	
			Eonychus, Crotonella, Tenuipalponychus
		Tetranychini	
			Brevinychus, Sonotetranychus, Mixonychus, Evertella, ***Panonychus****, Allonychus,* ***Schizotetranychus****, Yunonychus,* ***Yezonychus****, Neotetranychus, Acanthonychus, Mononychellus, Paltyetranychus,* ***Eotetranychus,*** *Palmanychus, Atrichoproctus Xinella,* ***Oligonychus,*** *Hellenychus,* ***Tetranychus, Amphitetranychus, Stigmaeopsis***[a]

Genera in bold are known in Japan
[a]Not listed in Bolland et al. (1998)

It is known that there are special food habits among the genera of the Tetranychini (Table 2.2). For example, the genus *Stigmaeopsis* only inhabits graminid plants, such that it is regarded as monophagous (Saito et al. 2004). On the other hand, the genus *Tetranychus* inhabits a wide variety of plants and is regarded as polyphagous (Gutierrez and Helle 1985). The genera *Panonychus* and *Oligonychus* are considered intermediate to *Tetranychus* and *Stigmaeopsis*. *Eotetranychus* and *Schizotetranychus* also occur on a wide variety of tree species. However, such an overview at the genus level is not always appropriate, if we focus on the individual species of each genus: the species of *Eotetranychus* and *Schizotetranychus* are almost all host specific, e.g., *Eotetranychus tiliarium* occurs only on *Tillia* spp., *Eotetranychus shii* only on chinquapins, *Schizotetranychus schizopus* only on willows, *S. cercidiphylli* only on katsura, *S. lespedezae* on Japanese wisteria, and so on. Thus, at the genus level, they have a wide variety of hosts, but at the species level, almost all species inhabit only a very narrow range of species of host plants. Therefore, there is little sense in saying that a spider mite genus has a certain type of food habit, as pointed out by

Table 2.2 Brief summary of food habits of genera in Tetranychidae (restricted to the genera distributed in Japan)

Genus	No. of species in the world	No. of species in Japan	Tree or grass	Evergreen or deciduous tree	Annual or perennial grass	Host specificity at species level
Aponychus	19	2	Grass	–	Perennial	High
Eutetranychus	32	1	Tree	Deciduous	–	High
Amphitetranychus	3	2	Tree	Deciduous	–	High
Eotetranychus	175	20	Tree	Both	–	High
Oligonychus	194	17	Both	Both	Perennial	Relatively high
Panonychus[a]	15	8[a]	Both	Both	Perennial	High
Schizotetranychus	114	6	Tree	Both	Perennial	High
Stigmaeopsis	7	5	Grass	–	Perennial	High
Tetranychus	135	10	Both	Deciduous	Both	Low[b]
Yezonychus	1	1	Grass	–	Perennial	High

[a]*Sasanychus* inhabiting perennial bamboo is considered to be synonym of *Panonychus* in accordance with Bolland et al. (1998)
[b]There are exceptions such as *Tetranychus ezoensis* Ehara.

Gutierrez and Helle (1985). Rather, such trends at the genus level mentioned above should be considered historical trends and should be reanalyzed from a phylogenetic point of view.

2.2 Life History and Host Plants

Saito (1979a) and Saito and Ueno (1979) showed that there is much variation in life history patterns among six spider mite species (see the typical three species in Fig. 2.1) and discussed that such diversity might be explained by the spatiotemporal stability of host plants. For example, a famous pest species, the two-spotted spider mite [*Tetranychus (Tt.) urticae*], has nearly the highest r_m (intrinsic rate of natural increase) among spider mites and *Stigmaeopsis (Schizotetranychus) longus* (long seta form of *Schizotetranychus celarius*, before Saito 1990a) shows very low r_m. Mitchell (1973) stated that the high r_m of *Tt. urticae* is a characteristic of colonizing species that are found on ephemeral plants (annual host plants) carrying a huge amount of resources, and that factor rapidly increases its population. This type of life history is advantageous if resources emerge randomly. A high potential for population increase must evolve in a species to use such habitats effectively and to produce sufficient dispersers to find new resources arising unexpectedly. On the other hand, in stable habitats with sufficient resources, so many competitors and natural enemies will coexist (or rather accumulate) and prey species have to spend much energy in the struggle for survival. Therefore, they should develop a kind of protection and/or defense mechanism instead of high reproductive rates as an evolutionary trade-off. In general, the spatiotemporal stability of host plants must be very low in the host plants of *Tt. urticae* in comparison with those of *St. longus*.

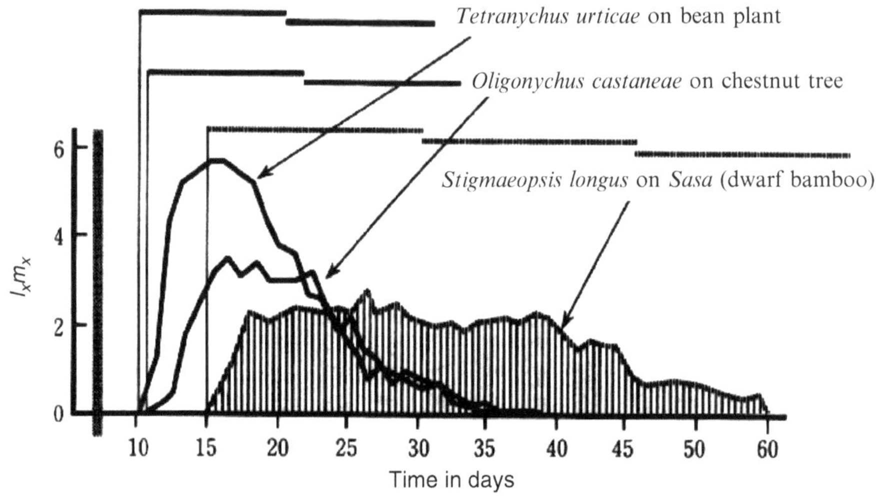

Fig. 2.1 Various life histories known in Tetranychinae explained from the spatiotemporal stability of host plants under 25°C, 50–60% r.h. (after Saito and Ueno 1979). Horizontal bar shows mean developmental period of each species.

Box 2.3 Problem in life history study of plant mites

The leaf disc method (sometimes called detached leaf culture), which has been used as a convenient method for rearing plant mites, involves a fatal problem for life history study. In this method, a detached leaf is placed on water-soaked cotton or polyurethane sponge in a Petri dish filled with water. Water surrounding the leaf effectively prevents mites on the area from escape, and it maintains the leaf alive for a long period. Therefore, it has been used as a simple applicable method to maintain mite colonies under laboratory conditions.

However, the temperature decrease by water evaporation from the cotton and Petri dish is sometimes not negligible. This effect is obviously serious under low-humidity conditions, such as in air-conditioned insectariums. Saito and Suzuki (1987) showed that the surface temperature (mites must be affected by this) of a detached leaf under 47% r.h. is about 21.5°C when air temperature is 24.8°C. Because the developmental time of *Tetranychus (Tt.) urticae* under 21.5°C is roughly corresponds to 1.5 times of that under 24.8°C, we cannot know the accurate developmental time, if there is a description of air temperature only without that of air humidity. Therefore, we should pay great attention to the temperature (and/or air humidity) of the microhabitat (i.e., leaf surface) when we conduct mite life history study by this method.

Thus, Saito and Ueno (1979) concluded that such a difference in life history parameters reflects host plant stability, as expected by Mitchell (1973). It was still an application of the well-known *r*/*K* selection or habitat template theories (McArthur and Wilson 1967; Southwood 1977) prevailing at that time in ecology, but they might have contributed to showing the importance of diversity in spider mite life history (Box 2.3). At present, we know that life histories vary depending upon phylogeny, antipredator adaptations (Sect. 3.6), and so on, but there can be no doubt that host plant stability has also fundamentally affected spider mite life histories (Sabelis 1985).

2.3 Life Type Diversity in Spider Mites

There is great variation in life type (or life style) between spider mite species (Fig. 2.2; Saito 1979b, 1985, 1995b). Life types are defined as "strategies by which mite species utilize silken threads on the leaves of their host plant." Even though the species studied are mostly restricted to those being distributed in Japan, I was able to show that spider mites have diversified over various plants and exhibit various kinds of life types on them. I have studied not only pest species but also various non-pest spider mite species to enable comparative ecological studies (Box 2.4).

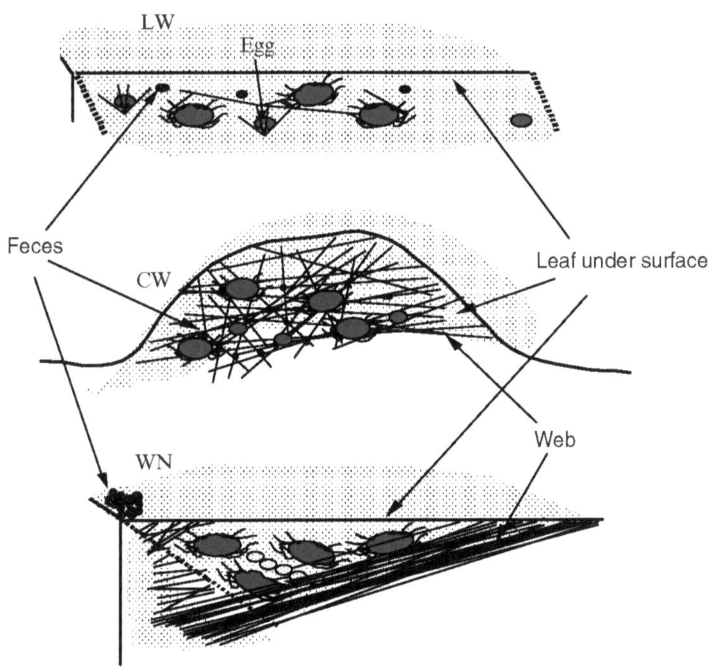

Fig. 2.2 Spider mite diversity in life type, little web type (*LW*), complicated web type (*CW*), and woven nest type (*WN*). Typical subtypes are shown later

Box 2.4

Spider mites had been thought to be somewhat homogeneous animals and were simply referred to as "red mites" in agricultural fields, because of their very tiny bodies with red color (not always so). When I started my biological study on spider mites, most researchers were thinking in such erroneous terms. One day Mr. Moriyama, who was my senior by 4 years, showed me several species of spider mites that occur on *Sasa*, which are the main cast of this book. This was when I first recognized the diversity of life style (life type) in spider mites. Therefore, observing them changed my research life and provided the opportunity that these spider mites became my long-term partners.

What Is a Life Type?

Let us describe several of the typical life types observed so far. The simplest life type, called LW (little web), was observed in *Aponychus firmianae*, which never produces threads during its lifetime (Fig. 2.3). This species belongs to Eurytetranychini, which is thought to be more ancient than Tetranychini. In the same tribe, we know another species, *Aponychus corpuzae*, which produces web covers on eggs (see Sect. 3.2); this is the simplest usage of silken threads in the Tetranychinae. *Panonychus citri* (and *Panonychus ulmi*) always pull out threads through their pedipalpi when walking (Sect. 2.3) and make "guy ropes" extending from their eggs to the leaf surface. These species are also considered to have the LW life type (Fig. 2.3). On the other hand, most *Tetranychus* species and some of the species of *Oligonychus*, *Schizotetranychus*, and *Eotetranychus* have another life type, the CW (nonsystematic and complicated web; Fig. 2.4) type. Saito (1977a) showed that *Tt. urticae* always pulls out threads when walking (Sect. 2.3). Because females of this species aggregate in leaf depressions (e.g., along leaf midribs), such threads accumulate in particular areas and form web structures. This process is considered to be a base of the CW life type. Furthermore, the mites tend to walk, to deposit eggs and feces (Hazan et al. 1974; Gerson 1985), and to be quiescent within the accumulated web. Thus, I hypothesized that these complicated webs must serve to avoid predation and hazards from adverse climatic conditions (Saito 1985, 1995b). Relative to the foregoing, it is known that leaf depressions on which CW webs are made are sometimes created by the spider mites themselves (Oku et al. 2006). These suggest that CW type webs are not only a result of mite movement with threads but also are produced for such functional reasons from aggregated life, as mentioned later (Sect. 2.3).

On the other hand, there are many species that build systematic structures by threads on the host leaf surface. I called this life type the woven nest (WN) type (Figs. 2.5, 2.6). *Eotetranychus suginamensis*, *E. shii*, and *Schizotetranychus brevisetosus* make web nests over depressions on leaves (on the underside), as shown in Fig. 2.6

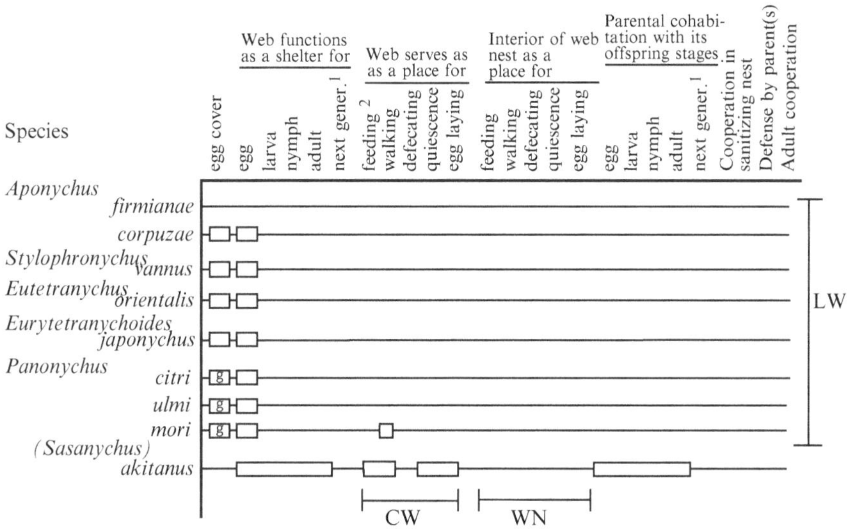

Fig. 2.3 Various life patterns observed in the genera of Tetranychidae. *Rectangles* indicate what stages of offspring usually share in the benefit of the trait or what traits were observed in the species. g, guy ropes; *1*, web serves for the next generation (*gener.*) (grandchildren); *2*, feeding within and under web; *LW*, little web type; *CW*, complicated web type; *WN*, woven nest type (after Saito 1995b)

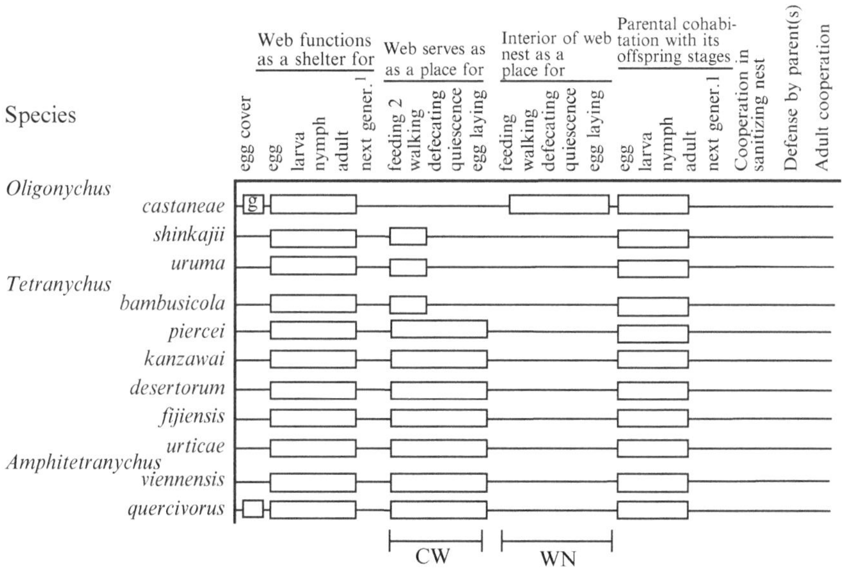

Fig. 2.4 Various life patterns observed in the genera of Tetranychidae. See Fig. 2.3 for explanations (after Saito 1995b)

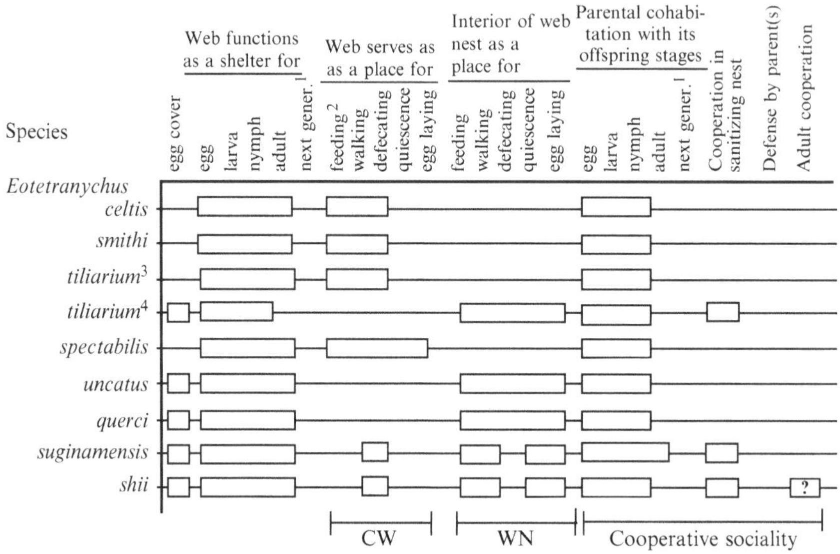

Fig. 2.5 Various social life patterns observed in *Eotetranychus*. *3*, On *Alnus hirsuta*; *4*, on *Alnus japonica*. See Fig. 2.3 for other explanations (after Saito 1995b)

(WN-s in Fig. 2.7). Individuals living under the web nests always deposit their feces on the outside nest roof (web), and their nest webs become coated with feces. Yokoyama (1932) observed this behavior and supposed that *E. suginamensis* fortifies its nest by using feces. However, I rather supposed that this behavior has the primary function of nest sanitation, because many species having nest-weaving habits show more or less different kinds of feces manipulation behaviors (Fig. 2.7; Saito 1995b). Among mites having the WN life type, the most sophisticated nesting behavior is seen in *Stigmaeopsis* species. This genus is novel to the Tetranychidae (Saito et al. 2004; the details are described in Sect. 3.5). All species belonging to this genus have a common life subtype, WN-c, which is characterized by their solid woven nests and special feces-depositing behavior. For example, in the largest nest builder, *Stigmaeopsis (St.) longus*, all nest members deposit their feces at a particular place in their nests (WN-c in Fig. 2.7), suggesting that there is a kind of cooperation in this species (Sect. 4.4). A single united nest is used for a long period, such that a big pile of feces within the nest is visible to the naked eye. Most species having the WN life type have a monophagous food habit and live on the leaves of evergreen trees and perennial grasses (e.g., *Sasa* and *Miscanthus*). Anyhow, the nest-weaving and fecal manipulation habits are two important features for defining the WN life type and also serve as criteria to define life subtypes (Saito 1995b).

We can thus recognize three basic life types, LW, CW, and WN, through web-building and feces manipulation patterns in Tetranychini. However, we also know that there are several life types that are now conveniently categorized into these three types Box. 2.5. In the CW type, *Panonychus* (*Sasanychus*) *akitanus* makes two to three layers of web over depressions on the host leaf undersurface (*Sasa senanensis*). Individual mites move on these layers and deposit eggs and feces there. Furthermore,

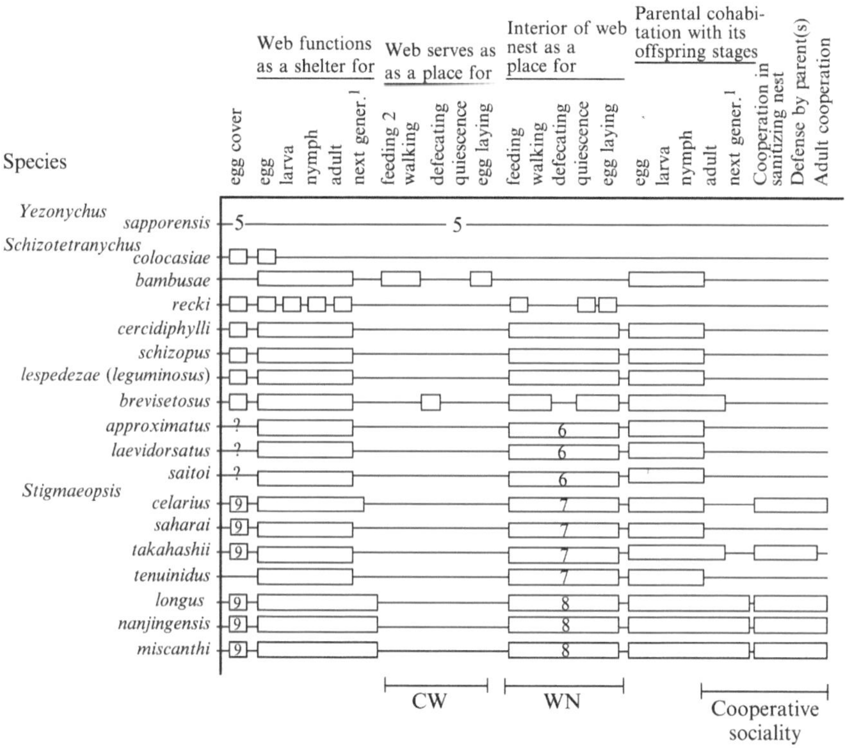

Fig. 2.6 Various social life patterns observed in three genera in Tetranychinae. *5*, A different means of egg protection and quiescent stage is known (see Sect. 3.3); *6*, feces are deposited outside of nest; *7*, outside of particular places near nest entrance; *8*, inside of particular place(s) near nest entrances; *9*, web mat is made under eggs. See Fig. 2.3 for other explanations (after Saito 1995b)

all quiescent stages rest on these layers and never on the leaf surface. Whether we have to classify this life pattern into CW or WN is another question, but this in itself may not serve any biological insight other than showing their diversity in life types. *Schizotetranychus recki* makes individual nests with leaf hairs (Sect. 3.4). *Eotetranychus asiaticus* shows another curious habit: many mite individuals live with CW webs in a gall of psyllid larvae that cover their gall openings (Fig. 2.8). Such a closed leaf hole must provide a very safe area from predator invasion, such that it is supposed that this habit has a kind of antipredatory function as well as functioning to protect mites from adverse climatic conditions (Saito, unpublished data).

As described above, such variation in leaf physical features is undoubtedly important for spider mites to establish their own life type by webs. There are many studies on competition between herbivorous mites, on interactions between herbivorous mites and their predators (Huffaker et al. 1963; Bernstein 1984; McMurtry 1985; Hoy 1985; Saito et al. 1996; Osakabe et al. 2006), and on indirect effects of plant structure (Fournier et al. 2003), which are more or less related to the web structures of these

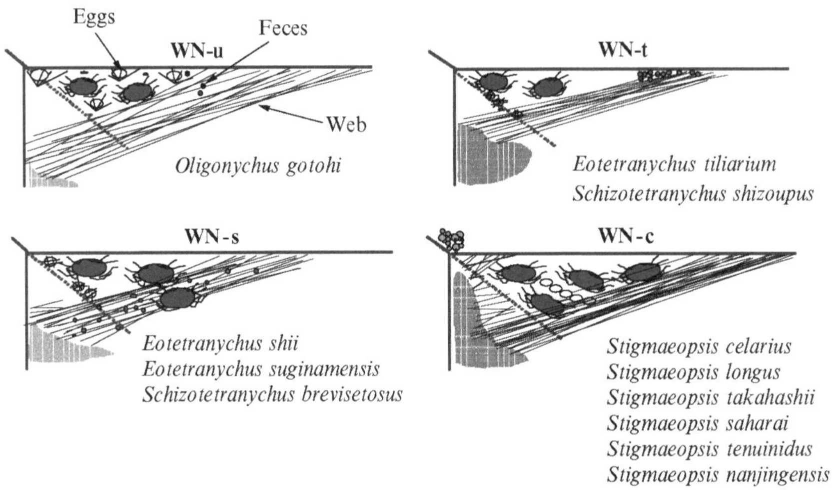

Fig. 2.7 Several subtypes in WN life type

Box 2.5

Recently I observed a curious life type on a Chinese bamboo plant that has quite dense hairs. These hairs force female mites to walk through a narrow space between the parallel-growing leaf hairs. The females of *Schizotetrany chus* sp. have very slender bodies, perhaps because of the physical structure of this habitat, and they move on the leaf surface in a column. Flat web nests are built on the tips of leaf hairs, as shown below. It is a little bit difficult to imagine the function of this web, but if the mites use the dense hairs surrounding their nest as a stockade (then there is a lower risk of being attacked from the lateral sides of nest) and protect themselves by the flat web from predators attacking the upper surface of the nest, the function of this kind of life type is understandable. However, I have still no information about such predator species.

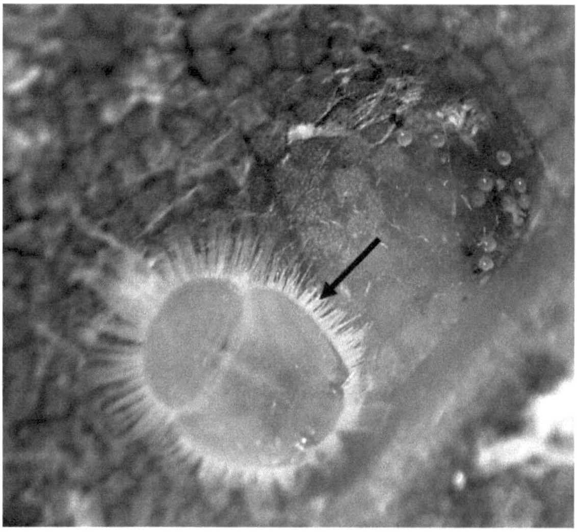

Fig. 2.8 Extraordinary life type known in *Eotetranychus asiaticus* Ehara living with CW life type in an open gall of *Trioza cinnamomi* (Boselli). The larval *T. cinnamomi* that had closed the gall was replaced (*arrow*). Is this symbiosis between spider mite and a psyllid (in Tanegashima, Japan)?

herbivorous mite species. Thus, we learned that spider mites have surely evolved under the selection pressures of host plant physical features, host plant phenology, and natural enemies (Chant 1959; Dosse 1961; Van de Vrie et al. 1972; Helle and Sabelis 1985; Sabelis and Bakker 1992; McMurtry and Croft 1997).

Supposed Function of LW and CW Life Types

From the foregoing discussion, we know that spider mites have very diverse lives that vary among species and among host plant species. If so, we have next to address basic problems about how and why spider mites evolved to produce silken threads and to have their own life types.

Before describing the function of each life type, it is useful to determine what the basic function of silken threads is. I previously introduced that *Panonychus (Pn.) citri* and *Tt. urticae* always produce silken threads when walking (Saito 1979a). It is thought that the name "spider mites" was originally given to these animals because several well-known pest species produce silken threads just as do spiders (Lindquist 1985). There are several taxa in the Acari other than Tetranychidae in which some species produce silken threads: Cheyletidae, Anystidae, Bdellidae, and so on (Walter and Proctor 1999). However, most of them only use the threads to make cocoon-like capsules when they are quiescent (Anystidae) or to make traps to capture prey (Bdellidae). Hence, the Tetranychini and Cheyletidae (see Sect. 4.6) are the only groups in the Acari that use silken threads so diversely throughout their lives.

Spider mites produce two thread lines from a pair of organs (called terminal eupathids) on the pedipalpi, as shown by the scanning electron microscope (Jeppson et al. 1975; Box 2.6). They never eject these silken threads but rather use ambulatory movement to pull silk out from the organs; i.e., they stick the tips of the silk threads from the silk production organs onto the substrate (leaf surface) and then silken threads are pulled out as they walk (Fig. 2.9; Mori and Saito 2006) or sometimes by their falling down.

Box 2.6

Before Jeppson et al. (1975), there were two opposing hypotheses about where spider mites produce silken threads; i.e., one has been mentioned in the text (Grandjean 1948), and the other is that a single thread is produced from the rostrum (Mills 1973). Because of the extreme thinness of silk threads, such a discrepancy in observation was inevitable. Saito (1977a) also observed that a *Tetranychus urticae* female produces a pair of threads simultaneously. From very closely located pairs of pedipalpi and silk-producing organs (see below), two silk threads readily stick to each other. Gerson (1979) estimated the thickness of *Tt. urticae* thread to be the diameter of a single thread is 0.03 μm and that of a wound thread is 0.06 μm. These measurement values are reasonable if there are a single thread and stuck (double) threads.

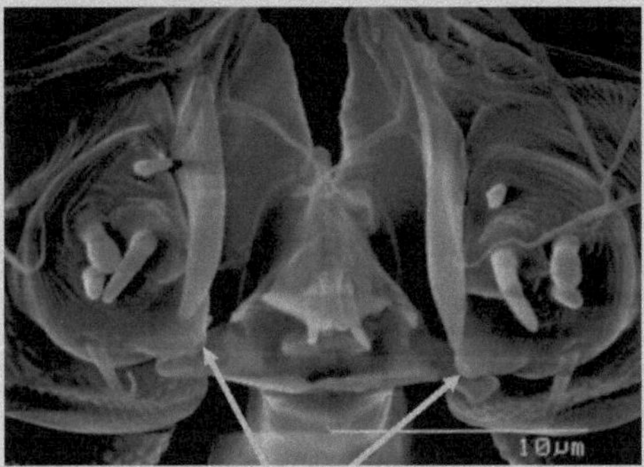

Terminal euphathids
(Silk producing organs)
of *St. miscanthi*

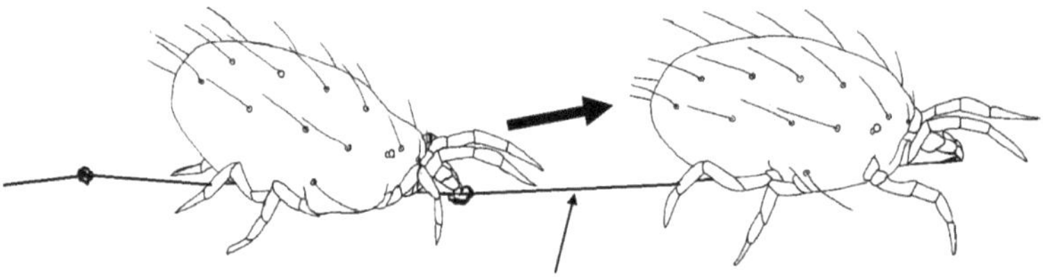

Fig. 2.9 How a spider mite draws out its silken threads

Box 2.7

Silk threads are almost invisible even under dissecting microscope. How did I idealize this method for the quantitative evaluation of webs? In the 1950s, in Japan, children had no toys such as they do today, and we often played in fields using homemade apparatus to capture insects. One such tool was made by circles of wire. Spider's webs were swept into these circles for use as adhesive sweeping nets. This memory, as well as the method used to evaluate walking velocity by Fransz (1974), provided the hint to use lattices with thin nylon threads. Of course, the minimum thickness of the available nylon thread was 50 μm, such that it was very laborious for spider mites, which are only 400 μm in body length, to move over such relatively big obstacles.

Saito (1977a) developed a method to quantitatively evaluate the amount of threads that *Pn. citri* and *Tt. urticae* produce. A fine mesh gauze (1 mm × 1 mm mesh size) consisting of thin artificial threads (30–50 μm thick) was made and placed on the leaf surface. Then, females of these two mite species were released on the leaf surface (facing downward) with the mesh. By use of an overhead projector, I first traced the walking paths of these mites, and then picked up the lattice net (with produced threads still attached). Under a microscope with dark phase contrast, the number of threads traversing each lattice (square) of the mesh was counted, and then the number of line segments traced and that of walking paths traversing each square of lattice were compared (Box 2.7). In this way, I showed that the number of walking paths traversing squares and that of threads produced on lattices are well correlated in a 1:1 manner (Fig. 2.10; Saito 1977a). This result shows that the females of these two species always pull silk threads out as they walk and it corresponds well to the mode of silk production.

In the process of these observations, I also observed a curious phenomenon with an inverted-type projector; i.e., *Tt. urticae* (and sometimes *Pn. citri*) while walking on the undersurface of the plant leaf (facing downward) occasionally lost their footing and fell (Saito 1977a). This observation was unbelievable for me because I had predicted that the spider mites living on the leaf undersurface (most spider mites live on the leaf undersurface, but some species, such as *Pn. citri*, *Pn. ulmi*, *Oligonychus castaneae*, … etc., tend to live on the upper surface) have sufficient adaptations to prevent them from falling. However, I simultaneously observed that the mites falling down immediately returned to the leaf surface. What happened in these mites is obvious. They always produce threads when walking, and they stick these threads to the leaf surface at certain intervals. When they lost their foothold, they were suspended by their threads from the leaf surface and could easily climb back to the leaf along the silken threads. Such a phenomenon is sometimes observed just after acaricide spraying (Fig. 2.11; Gerson and Aronowitz 1981). The mites are paralyzed by the pesticide, lose their footing, and are suspended by their threads.

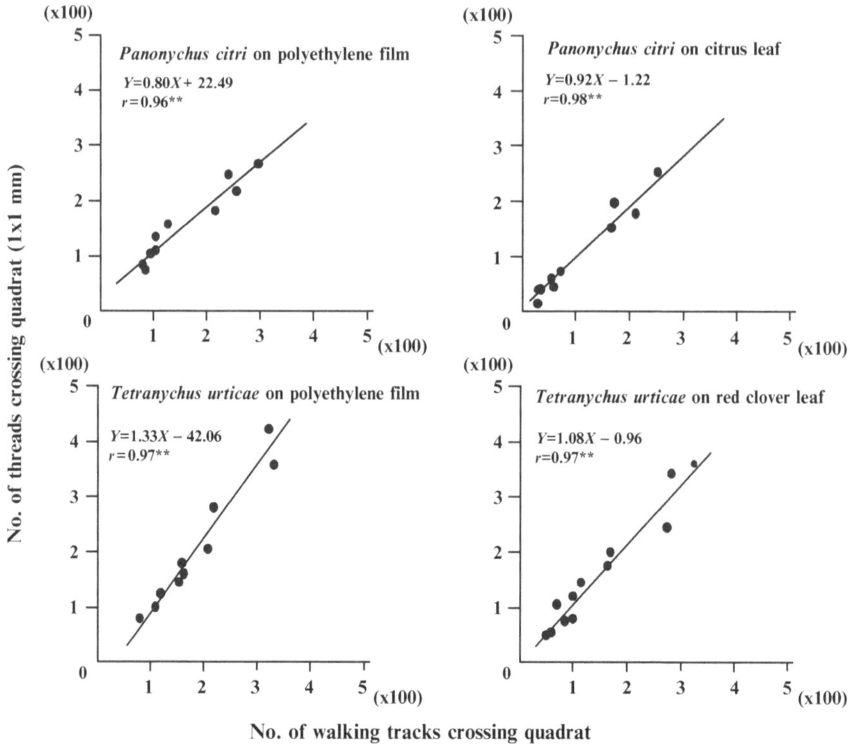

Fig. 2.10 There is almost a 1:1 correspondence between threads and walking tracks in these two spider mite species (after Saito 1977a)

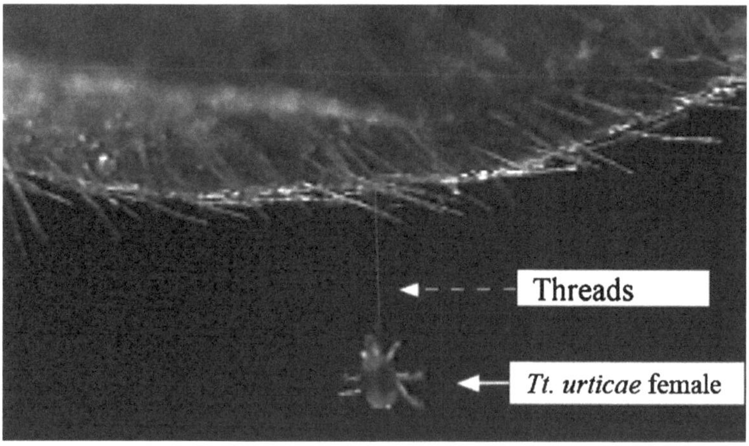

Fig. 2.11 *Tetranychus (Tt.) urticae* female suspended from leaf surface by threads. Because the threads are invisible, a *white line* was added on the original photograph to show the thread (from VTR made by Nihon Nouyaku Co. Ltd)

Therefore, Saito (1977a) concluded that one of the key – and perhaps primary – functions of silken threads is as "lifelines" that enable mites residing on the host leaf undersurface to avoid the disadvantage of losing contact with the food plant. Such a function of threads must be regarded as a convergence of both spider and some larval insect threads. In relationship to the foregoing, Yano (2008) investigated the role of trails in collective and solitary behaviors of *Tt. urticae* females. He found that dispersing females simply follow the trails left by preceding females, then they disperse collectively. Although he did not ascertain what constitutes a trail, he discussed the possibility of a pheromone and/or threads. As Penman and Cone (1972) and Oku et al. (2005) reported the importance of web plus female sex pheromone, which simulated male searching behavior for females (quiescent deutonymphs), I believe that Yano's (2008) discovery is closely related with the threads (and pheromone attached to them) produced by female walking.

On the other hand, spider mite threads are not always lifelines to fix them on to the host plant leaves but can be apparatuses for dispersal from their residence (Helle and Sabelis 1985). Fleschner et al. (1956) reported such a paradox in their paper titled "Air drift of spider mites." They reported that among spider mites, several species disperse with silken threads (ballooning) from leaves when their host plants are severely damaged. This behavior is thought to be resemble the popular dispersal behavior of spiders, i.e., gossamer. Furuhashi and Nishino (1978) observed that the females of *Pn. citri* disperse from tree to tree by drift (extending threads) in calm winds. We also observed that many females of *Oligonychus castaneae* (*ununguis*) suspend by threads from heavily infested chestnut leaves and gradually extend threads. By swinging from leaf to leaf by the wind, they disperse and colonize new leaves (Wanibuchi and Saito 1983). Thus, spider mite threads have at least two opposite functions: "lifelines" when residing on leaves and tools for moving when dispersing from plant to plant. Occasionally, we were fortunate enough to see that both sides of the same coin apply under natural conditions.

In the manner mentioned above, silken threads are indispensable for many spider mite species. Such threads sometimes form web structures on the leaf surface, i.e., the webs of CW and WN life types. The phenomenon that attracted our attention was the huge webs (CW) of *Tetranychus* spp. (e.g., *T. lintearius*; plate 1B in Helle and Sabelis 1985) and *Tt. urticae* (Fig. 2.12), which almost entirely cover the host plant (Box 2.8). Why they should build such a web wrapping over almost withered (and heavily damaged) plants is a difficult question, because if webs are primarily produced to protect mites from predators and/or adverse climate conditions, they should first build such webs just after host plant colonization (as observed in the web nests of the WN life type; see later). Thus, I concluded that such huge webs are primarily a result of mite infestation, in other words, "by-products." After revealing that the threads function as lifelines, it is understandable to consider such web accumulation as the result of walking by a tremendous number of mites to search for food within deteriorated host plants.

On the other hand, the CW type web might have subsidiary functions in several spider mite species. It is known that there are two types of predator species, those that can intrude into CW webs and those that cannot (Sabelis and Bakker 1992;

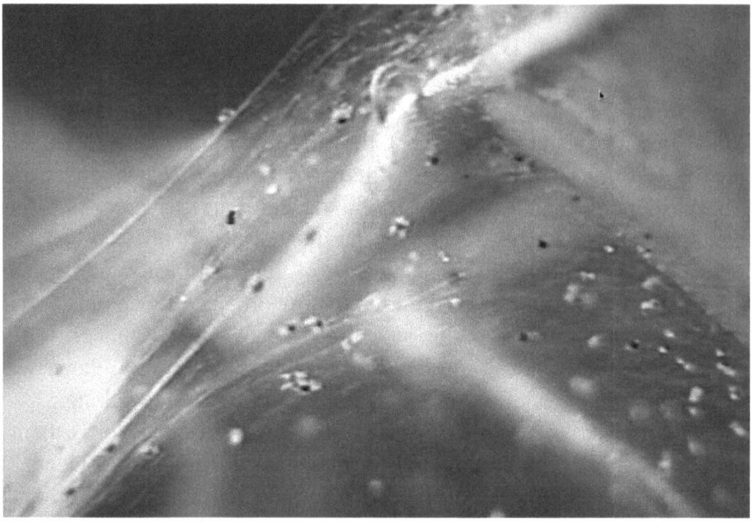

Fig. 2.12 Tent-like webbing produced after a heavy infestation of *Tt. urticae* on red clover (from VTR made by Nihon Nouyaku Co. Ltd)

Box 2.8

My first interest as a spider mite researcher was in this huge web. I was a graduate student of Prof. H. Mori, who had studied the biological control of the two-spotted spider mite by phytoseiid predators. The first theme that he provided me was a search for alternative food for phytoseiids. To keep phytoseiids in the laboratory, the rearing of spider mites as prey in a greenhouse was important work, but I often failed at it when the prey population overshot. As a result of this overshoot, all host plant leaves withered and tremendous mite webs covered entire plants as well as pots. The reason why spider mites produce such huge webs after they had exploited their host plants thus became a stepping stone to my web research.

McMurtry and Croft 1997), such that the webs sometimes function to protect spider mites from predation (Sect. 2.5). In several *Tetranychus* and *Oligonychus* species having a CW-u life type, larvae and nymphs enter quiescence on the web (see Fig. 2.4; note that the quiescent stage is as long as each active immature stage). Oku et al. (2003) reported that *Tetranychus kanzawai* deutonymphs more frequently enter quiescent stages (teleiochrysalis) on the web than on the leaf surface, when they (of course before becoming quiescent) are exposed to a phytoseiid predator, *Neoseiulus womersleyi*. Such a behavior suggests that quiescence on the web is safer from predator attack than quiescence on the leaf surface.

Furthermore, the behavior of depositing feces on the CW web, known in many *Tetranychus* species, is thought to have a function of manipulating the harmful effects of feces (Saito 1995b). When we experimentally reared spider mites having a LW life type (they live sparsely and seldom make webs), such as *Pn. citri* and *Aponychus (Ap.) corpuzae*, at a unusually high densities in narrow leaf areas, many individuals became trapped on the leaf surface by disorderly deposited feces. Such observations suggest that wet and sticky feces must be harmful if they are densely deposited on the leaf surface. Feces deposited on the web must dry more easily than that deposited on the leaf surface, such that depositing feces on the web is thought to be less disadvantageous to these spider mites (the importance of feces-depositing behavior is addressed later). Furthermore, the web may effectively repel various particles, such as bacteria, fungal spores, and dust, from outside, because it has been empirically known that plants growing in dusty areas are frequently damaged by spider mites. As such, we now have learned that spider mite webs (threads) have multi functions.

The behavior of depositing feces on the web was used by Hazan et al. (1974) to observe the webbing behavior of *Tt. urticae (cinnabarinus)*. They noticed that *Tt. urticae* females tend to deposit their feces on webs and supposed that the number of feces (only for blackish grains other than pale yellowish ones) must be correlated with the amount (density) of web if they deposit feces randomly on webs and leaf surfaces. Then, Hazan et al. (1974) were the first to evaluate spider mite webs quantitatively by counting the numbers of feces on the web and on the leaf surface. Just after their studies, Saito (1977a) developed a method to count web threads directly, as already described, and showed that the amount of webbing measured by Hazan et al. (1974) is not the amount of web but rather can be regarded as the density of the web. In any case, they showed that *Tt. urticae* produces much more web under low-humidity and high-temperature conditions (Hazan et al. 1974; Saito 1977b). If the activity of mites tends to increase with temperature under low humidity, then these findings correspond well to the fact that the mites always produce threads when walking.

The webs of CW life types sometimes have an effect on competition between spider mite species. It has long been empirically established in apple orchards that *Pn. ulmi* is abundant in the early season and *Tt. urticae* replaces *Pn. ulmi* in the late season (H. Mori, personal communication, 1977). Morimoto et al. (2006) and Osakabe et al. (2006) showed that *Tt. urticae* excludes *Pn. ulmi* under both experimental and field conditions. They suggest that the abundant web of the former is the likely reason for the displacement of the latter. As mentioned earlier, *Tt. urticae* has a CW life type whereas *Pn. ulmi* has a LW type. This finding suggests that the former has a high ability to move on a dense complex web but the latter does not. The ability to move under, in, and on the web is thought to be closely related to the structure of the walking apparatus (ambulacrum): The empodium (organ at the tip of legs) of *Tt. urticae* consists of three pairs of fine hairs (arrows in Fig. 2.13), and the *Pn. ulmi* empodium consists of one big claw (arrow in Fig. 2.13) with three pairs of fine hairs. As shown by Gutierrez and Helle (1985), *Tt. urticae* can easily walk on threads with its three pairs of empodial hairs opening radially, because such a feature may enable mite legs to take on thin threads. On the other hand, the hooked claws of *Pn. ulmi* and *Stigmaeopsis (St.) longus* (arrows in Fig. 2.13) are

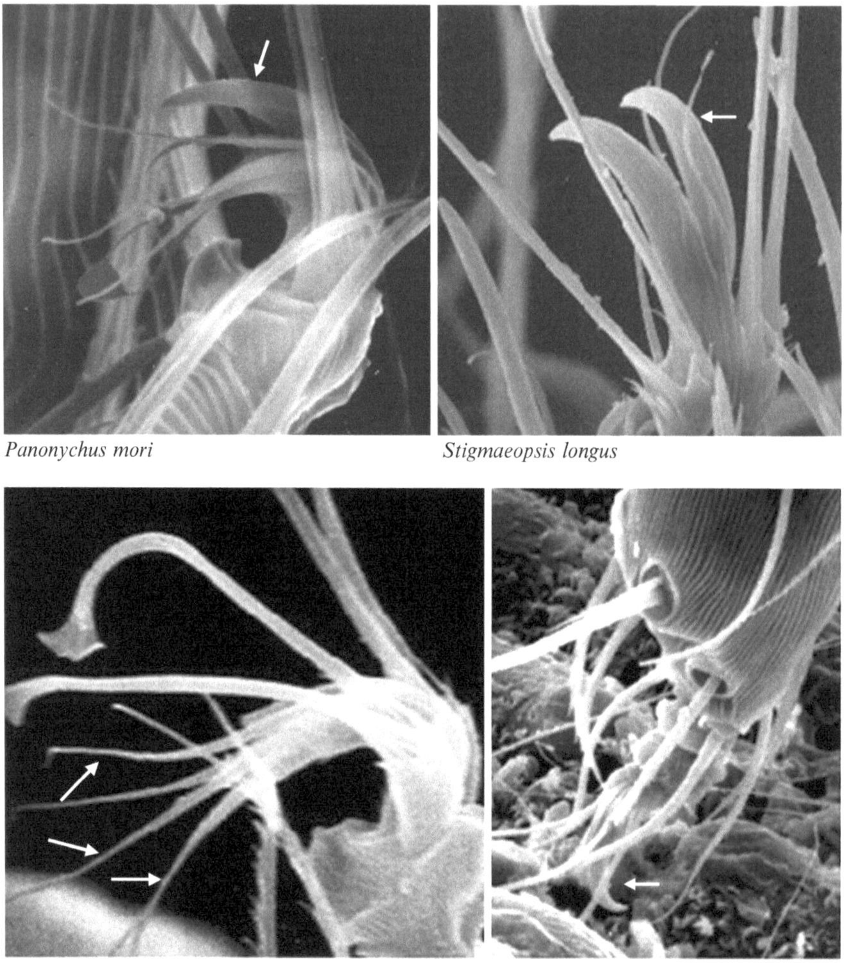

Panonychus mori *Stigmaeopsis longus*

Tetranychus urticae *Stigmaeopsis longus* grasping leaf surface

Fig. 2.13 Empodial claw variation among spider mite genera

expected to be functional for grasping the leaf substrate but must get hooked on threads when walking on a *Tt. urticae* web. Actually, *Pn. citri* frequently becomes entangled when it walks on *Tt. urticae* webs. Therefore, *Pn. ulmi* may have difficulty living on leaves where *Tt. urticae* CW webs have accumulated, thus opening the way for its displacement by the latter, as shown by Osakabe et al. (2006).

However, although there are several reports about the functions of CW and WN life types, they are restricted to the species that are serious pests in agriculture. If we turn our eyes to natural vegetation, there are many spider mite species that build woven nests on leaf surfaces, i.e., species having WN life types. Shortly I address in detail the functions of this life type for the species inhabiting *Sasa senanensis* (Chap. 3). But before that, it is necessary to discuss spider mite diversity a little more.

2.4 Diversity of Physiological Adaptation in Spider Mites

There is another important trait in the spider mite life cycle, i.e., the diapause that enables them to overwinter during the severe winters of cold and temperate zones. Because spider mites include so many agricultural pest species in temperate zones, many researchers have studied their diapause attributes.

Variation in Overwintering Stage

There is variation in overwintering (in this book, I often refer to "overwinter" instead of "diapause," because diapause is a physiologically definable trait and must be used only after its attributes are experimentally determined) stage among tetranychid species. As shown in Table 2.3, there are at least three overwintering types in Tetranychinae. Overwintering in the egg stage is known in *Panonychus*,

Table 2.3 Overwintering stages in Tetranychinae

Species	Overwintering stage(s)	Sources
Amphitetranychus viennensis (McGregor)	Adult female	Veerman (1985)
Aponychus corpuzae Rimando	Adult female	Gotoh and Shida (2007)
Eotetranychus carpini (Oudemans)	Adult female	Veerman (1985)
Eotetranychus celtis Ehara	Adult female	Ehara and Shinkaji (1996)
Eotetranychus hicoriae (McGregor)	Adult female	Veerman (1985)
Eotetranychus hirsti Pritchard & Baker	Adult female	Veerman (1985)
Eotetranychus matthyssei Reeves	Adult female	Veerman (1985)
Eotetranychus populi (Koch)	Adult female	Veerman (1985)
Eotetranychus pruni (Oudemans)	Adult female	Veerman (1985)
Eotetranychus querci Reeves	Adult female	Veerman (1985)
Eotetranychus smithi Pritchard & Baker	Egg and adult female?	Ehara (1993)
Eotetranychus spectabilis Ehara	Adult female	Niino and Sakagami[a]
Eotetranychus suginamensis (Yokoyama)	Adult female	Ehara and Shinkaji (1996)
Eotetranychus tiliarium (Hermann)	Adult female	Veerman (1985)
Eotetranychus uchidai Ehara	Adult female	Gotoh (1984)
Eotetranychus uncatus Garman	Adult female	Veerman (1985), Gotoh (1987a)
Eotetranychus willametti (McGregor)	Adult female	Veerman (1985)
Eotetranychus yumensis (McGregor)	Adult female	Veerman (1985)
Eurytetranychs buxi (Garman)	Egg	Veerman (1985)
Neotetranychus runi Trägårdh	Adult female	Veerman (1985)
Oligonychus castaneae Ehara & Gotoh	Egg	Niino and Sakagami[a]
Oligonychus ilicis (McGregor)	Egg	Veerman (1985)
Oligonychus newcomeri (McGregor)	Egg	Veerman (1985)
Oligonychus pratensis (Banks)	Adult female	Veerman (1985)
Oligonychus rubicandus Ehara	Adult females	Niino and Sakagami[a]
Oligonychus ununguis (Jacobi)	Egg	Veerman (1985)

(continued)

Table 2.3 (continued)

Species	Overwintering stage(s)	Sources
Oligonychus yothersi (McGregor)	Egg	Veerman (1985)
Panonychus akitanus (Ehara)	Egg and adult female	Gotoh (1986)
Panonychus bambusicola Ehara & Gotoh	Egg	Gotoh (1987b)
Panonychus mori Yokoyama	Egg	Fujimoto and Takafuji (1993)
Panonychus thelytokus Ehara & Gotoh	Egg	Ehara and Shinkaji (1996)
Panonychus pusillus Ehara & Gotoh	Egg	Gotoh (1986)
Panonychus ulmi (Koch)	Egg	Veerman (1985)
Platytetranychus multifigituli (Ewing)	Adult female	Veerman (1985)
Schizotetranychus brevisetosus Ehara	Egg and adult female	Saito[a]
Schizotetranychus cercidiphylli Ehara	Egg	Gotoh (1989)
Schizotetranychus lespedezae Begljarov & Mitrofanov	Egg	Gotoh (1989)
Schizotetranychus recki Ehara	Egg	Gotoh and Shida (2007)
Schizotetranychus schizopus Zacher	Egg	Gotoh (1984)
Stigmaeopsis longus (Saito)	Adult female	Saito et al. (2005)
Stigmaeopsis miscanthi (Saito)	Adult female	Saito et al. (2005)
Stigmaeopsis saharai Saito & Mori	Adult female	Saito[a]
Stigmaeopsis takahashii Saito & Mori	Adult female	Saito[a]
Tetranychus canadensis (McGregor)	Adult female	Veerman (1985)
Tetranychus desertorum Banks	Adult female	Veerman (1985)
Tetranychus ezoensis Ehara	Adult female	Gotoh (1984)
Tetranychus kanzawai Kishida	Adult female	Veerman (1985)
Tetranychus mcdanieli McGregor	Adult female	Veerman (1985)
Tetranychus pacificus McGregor	Adult female	Veerman (1985)
Tetranychus schoenei McGregor	Adult female	Veerman (1985)
Tetranychus turkestani (Ugarov & Nikolski)	Adult female	Veerman (1985)
Tetranychus urticae Koch	Adult female	Veerman (1985)
Yezonychus sapporensis Ehara	Egg and adult female	Ehara (1993)

Here I regarded hibernal diapausing stage in Veerman (1985) as overwintering stage
[a] Unpublished data

Oligonychus, and *Schizotetranychus* species. Overwintering in the adult female phase is known in *Tetranychus*, *Eotetranychus*, *Oligonychus*, and *Stigmaeopsis* species. Although most species belonging to each genus have common overwintering stages, there are some exceptions in *Oligonychus*, *Yezonychus*, and *Panonychus* (*Sasanychus*) in which several species overwinter at different stages. For example, there are two types, either egg or adult females, in *Oligonychus* (see Table 2.3). Furthermore, we know that *Yezonychus sapporensis*, *Panonychus akitanus*, and *Schizotetranychus brevisetosus* overwinter in two stages, egg and adult female (Gotoh 1986; Gotoh and Shida 2007; Saito, unpublished data).

Why there are such differences in overwintering stages between spider mite genera and between species in a particular genus has been sometimes explained by the host phenology hypothesis (Gutierrez and Helle 1985). Among *Oligonychus* species, most of which inhabit trees, the grass-inhabiting species *Oligonychus pratensis* (Banks) and *O. rubicandus* undergo diapause at the adult (female) stage (see Table 2.3; Jeppson et al. 1975). Furthermore, *Tetranychus* species, most of which are grass inhabiting and very polyphagous, also overwinter as adult females (Table 2.3). Therefore, Gutierrez and Helle (1985) suggested that an advantage of the female diapause type mainly observed in grass inhabitants may be that after reactivation the adult females have better chances of survival and a broader dispersal area, compared with the larvae hatching from diapausing eggs. In contrast, in species living on trees, overwintering eggs deposited on the bases of their own host branches do not need any such dispersal ability because the larvae hatching from the eggs can easily reach new spring leaves. Because most overwintering eggs (diapaused) have thick shells that guarantee higher tolerance to cold weather and desiccation than summer eggs (Lees 1953), it is advantageous for species living on trees to overwinter at the egg stage (if larval hatching and leaf foliation is sufficiently synchronized). This host phenology hypothesis in the diapause stage may also partly apply to most spider mites species distributed in Japan (Table 2.3). However, *Stigmaeopsis* species overwinter at the adult female stage on *Sasa* plants, most of which never defoliate, such that a single leaf can live at least for 2 years (Saito, personal observation). Furthermore, *Eotetranychus* species that hibernate at the adult female stage (Table 2.3) cannot be explained, because most of these are tree inhabitants. Therefore, the host plant phenology hypothesis is applicable to several genera, but not to these two genera.

Diapause

So far I have avoided using the term diapause because, with the exception of several species (Lees 1953; Veerman 1985; Gotoh 1989; Ito and Saito 2006; Saito et al. 2005; Kawakami et al. 2008), we have little evidence of whether the overwintering stages of spider mites are truly induced diapause. Diapause-induced females stop oviposition, and some of them change their body color into carmine red or yellow. However, diapause induction in females is not always accompanied with a body color change (Saito et al. 2005; Box 2.9). Therefore, to determine diapause induction, we have to confirm the cessation of reproduction (reproductive diapause) for females or nonhatching of eggs (egg diapause) without a cold period, such that the data for species in Table 2.3 are not always sufficient. Precisely how short a daylength is required to induce diapause (critical photoperiod) and how long a period they have to be kept under cold conditions before diapause termination varies between species and between populations distributed in different latitudes were studied mainly for *Tt. urticae* (Veerman 1977; Koveos et al. 1999) and *Pn. ulmi* (Lees 1953). Gotoh's (1989) study on the diapause of several spider mite species inhabiting deciduous trees revealed that the occurrence of diapause-induced females that deposit diapause eggs is strongly influenced by the quality of food (plant) on which the females developed. Entering diapause must be advantageous

Box 2.9

It had been believed that diapausing female spider mites always changed their body color thus, body color change is a useful criterion for diapause induction. In most studies examining female diapause in spider mites, changes in female body coloration to bright orange (or sometimes yellow) are used as the sole criterion of diapause induction (Veerman 1985; Mochizuki and Takafuji 1996). In this study, we discovered an important fact: that *Stigmaeopsis* females induced to diapause (as measured by the cessation of reproduction) do not change their body color under experimental conditions (Saito et al. 2005). Therefore, these observations indicated that this criterion is not applicable to the diapause of *St. miscanthi* and *St. longus* females (nor to that of *Yezonychus sapporensis*; K. Ito, personal communication, 2002). On the other hand, under field conditions, we observed bright orange females in midwinter, suggesting that the body coloration of overwintering *St. miscanthi* females does not always synchronize with diapause induction. Therefore, we must consider that reproductive diapause is different from the acquisition of cold-hardiness that is generally known to be related to body coloration (Leather et al. 1993; Saito et al. 2005). Because the overwintering females of *St. miscanthi* and *St. longus* show a bright orange color in the field, identifying the factors that change the body color of diapause females is an open problem for future study.

for mites developing on low-quality leaves because no recovery of leaf quality can be expected in deciduous trees within the year. The fact that host difference greatly influenced female diapause incidence in *Tetranychus kanzawai* may also show that the host phenology through food quality influences diapause performance (Ito and Saito 2006). Furthermore, Goka and Takafuji (1991) reported that the diapause induction rate in tree-inhabiting *Tt. urticae* populations shows clinal variation with latitude in Japan, whereas herb-inhabiting *Tt. urticae* show a low diapause induction rate and no such trend. This finding means that differences in host quality and in host species partly explain the variation in diapause traits of spider mites.

On the other hand, Goka and Takafuji (1990) studied the genetics of diapause attributes of *Tt. urticae*. They suggested that the diapause of this species is determined by two factors: one is represented by multi-locus nuclear genes and the other by cytoplasmic factors. Furthermore, Yano and Takafuji (2002) reported that the diapause intensity of *Tt. urticae* is correlated with its dispersal ability: they were able to artificially select mites that had either a high or a low dispersal capacity. Then they observed the diapause incidence rate for these selected lineages and showed that mites with a genetically high dispersal capacity have a higher diapause incidence and a lower performance compared to mites with a low dispersal capacity. Their results suggest that differential dispersal capacity is associated with contrasting life history patterns as a result of natural selection, which may mean that the tactics to escape adverse environmental conditions at both spatial and temporal scales are genetically correlated (Goka 2001).

Although several important discoveries were made for certain species, overall information about diapause diversity among spider mite species is too sparse to discuss further.

2.5 Diversity of Plant Mites Other Than Spider Mites

In her famous book *Silent Spring*, Carson (1962) allegorically described a relationship between pesticides, pest spider mites, and their predators: After spraying of chemical pesticides, spider mites that have high tolerance to such chemicals can cease the web production necessary to protect themselves against predators. Then, the pest mites can increase their populations tremendously. We now know that *Tetranychus* and *Panonychus* spider mites, to which she must have been referring in her book, cannot cease silk production (Sect. 2.3), but her suggestion that the web has some function to prevent predation is true. We know various species of predators occur on plants. Predacious mites belonging to the families Phytoseiidae and Stigmaeidae are common on plants and are considered to be the other main components of plant mites (see Sect. 1.3). In Japan, 74 species of phytoseiid mites and six species of stigmaeid mites are known (Ehara et al. 1994; Ehara and Shinkaji 1996), and the number of species in the former is very similar to that of spider mite species, suggesting their evolutionary relationship.

Phytoseiids

Chant (1959) first stressed the importance of phytoseiid mites as natural enemies of spider mite species by his own experiments as well as field observations. His report must be considered a classic among studies about predator–prey interactions. After his work, several important studies on phytoseiids were made by Dosse (1961), McMurtry and Johnson (1966), and Van de Vrie and Boersma (1970). After their pioneer studies, many studies on the predatory efficiency of phytoseiids have been conducted for various species (Helle and Sabelis 1985), although most of them were concerned with agriculturally important species such as *Phytoseiulus (Pl.) persimilis* (Fig. 2.14), *Amblyseius orientalis*, *Neoseiulus* (*Amblyseius*) *cucumeris*, and *Neoseiulus womersleyi* (McMurtry and Croft 1997). One of these, *Pl. persimilis*, is famous worldwide as a natural enemy that effectively suppresses *Tetranychus* mites, which are serious pests on greenhouse vegetables, fruits, and crops, and this species is now mass produced and supplied by several biological control companies.

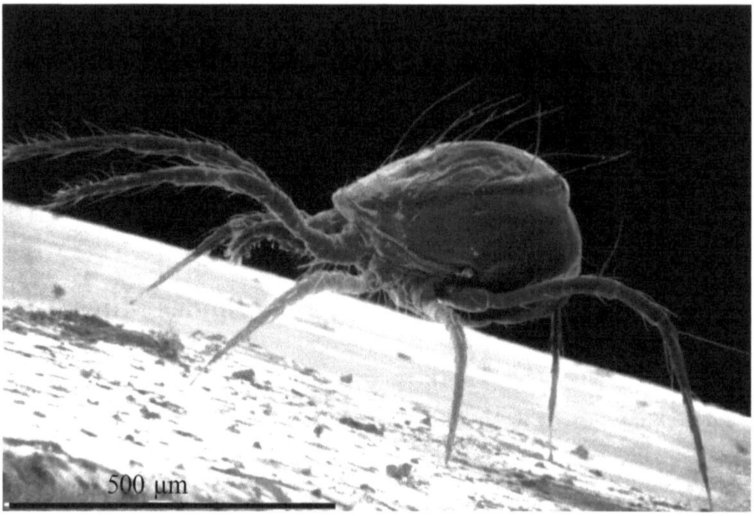

Fig. 2.14 *Phytoseiulus persimilis* Athias-Henriot, the most well known phytoseiid species in the world (scanning electron microscopy, SEM)

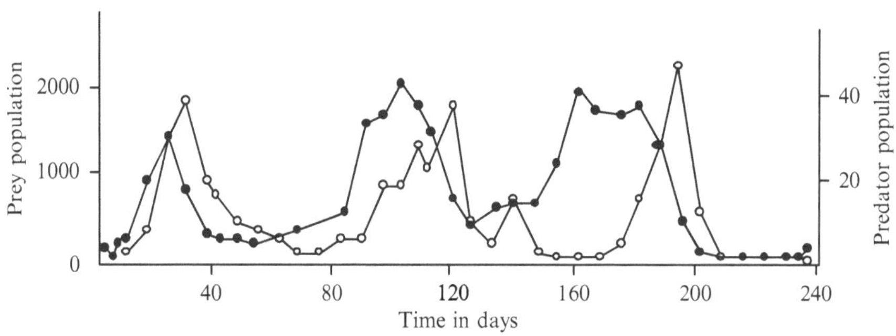

Fig. 2.15 Stable predator (*Typhlodromus occidentalis*, open circle)–prey (*Eotetranychus sexmaculatus*, black circle) interaction under complex (heterogeneous) environment conditions (redrawn from Huffaker 1958)

On the other hand, phytoseiid mites were also used as model animals to observe predator–prey interactions from the aspect of animal ecology. Huffaker (1958) indicated that there is a condition under which predator and prey can maintain stable populations, as in a Lotka–Volterra model (Lotka 1925; Volterra 1926), with some oscillation in a single system (Fig. 2.15). It is well known that most predator–prey systems are unstable (Kuno 1987), because there was a serious limitation of assumption (strict density dependence) in the Lotka–Volterra model. Therefore, Huffaker's experimental results showing that *Eotetranychus sexmaculatus* and *Typhlodromus occidentalis* populations can persist for long periods within a single system were extraordinary, and up to the present day these findings are frequently referred to in many biology texts (e.g., Begon et al. 1999). The reasons why his system could persist for so long are possibly the heterogeneity of habitat conditions (in the experimental system) under which prey (but not predator) can easily move between structured sub-patches within habitats. Anyway, this study on the interactions between spider mites and phytoseiid mites has contributed so much to showing "mites" as useful organisms for testing basic hypotheses in ecology.

Let me further discuss the diversity in predatory mites. McMurtry and Johnson (1966) first reported that there is variation of response to web structures between several phytoseiid mites: *Amblyseius hibisci* finds it difficult to penetrate *Oligonychus punicae* webs, but *Pl. persimilis* and *Typhlodromus (Tp.) occidentalis* can easily do so, such that the former species cannot be an effective predator of this prey species. McMurtry and Croft (1997) reported a comprehensive review about phytoseiid life styles in relationship to their roles in biological control. They categorized phytoseiids into four types based on food habits and related biological and morphological traits: type I, specialized predators of *Tetranychus* spp.; type II, selective predators of tetranychid mites; type III, generalist predators; and type IV, specialized pollen feeders/generalist predators (Box 2.10). Why such types are recognized in phytoseiids may be closely related to the ability to handle spider mite threads (web and nest). Type I phytoseiids must have a strong ability to penetrate CW type webs and effectively feed on spider mites within webs. Type II phytoseiids may include two subgroups: one feeds on LW type prey and the other feeds on WN type prey. Aponte and McMurtry (1997) reported how *Oligonychus perseae*, a serious pest of avocado, builds web nests on avocado leaves, thus showing a WN life type. Montserrat et al. (2008) recently showed how the predatory mite *Neoseiulus californicus* rips and

Box 2.10

My first study on mites concerned pollen as an alternative food for Japanese phytoseiids. To me this theme looked very elegant because collecting pollen is just the same as bees visiting flowers. However, I immediately noticed it was so laborious because it was impossible to conserve pollen in a fresh state on the detached leaf cultures (Box 2.3) usually used for mite rearing. This method consists of detached leaflets (or a piece of plastic) on wet cotton in a Petri dish with water. Pollen put on such a leaf immediately soaks up water and becomes unpalatable to phytoseiids. To avoid this, I developed a glass tube method in which water is sealed by wax and pollen is placed on a small shelf (Saito and Mori 1975). By this method, a little water coming through the slit between the wax seals is available for phytoseiids and pollen on the shelf set apart from the wax never becomes wet. This method is very useful to observe phytoseiid food habits as well as the tolerance of starvation. Chittenden and Saito (2001) used it for their larval feeding type experiments.

then penetrates the densely woven nests (WN type) of *O. perseae*. They observed triangular openings after this predator penetrated the nests of *O. perseae*, such that they supposed the phytoseiid can somehow rip nest threads. Very recently Shimoda et al. (2008) observed how *Pl. persimilis*, *Neoseiulus womersleyi*, and *N. californicus* can all cut the threads of a CW web using their chelicerae and palpi. On the other hand, *Typhlodromus (Tp.) bambusae*, a specific predator of *Stigmaeopsis* species, can easily intrude into woven prey nests using the nest entrances without ripping threads. Furthermore, the result of Mori and Saito (2004) suggest no mite predators are able to destroy the dense nest web of *Stigmaeopsis* species (Sect. 3.5). These studies suggest that there is much variation both in the strength of spider mite threads and in the ways phytoseiids manipulate them.

In relation to the foregoing, Sabelis and Bakker (1992) supposed that the long dorsal setae of some phytoseiids must be an adaptation to manipulate prey webs, because species that adapt well to webs commonly have long dorsal setae, i.e., the so-called "chaetotaxy hypothesis." *Pl. persimilis* has long dorsal setae (see Fig. 2.14). In Japan, *N. womersleyi*, an effective predator of *Tetranychus* species (CW type), also has long dorsal setae, as does *Tp. bambusae* (Fig. 2.16), a specific predator of *Stigmaeopsis* species that build dense-woven nests (WN type; Fig. 2.7). Therefore, their hypothesis may have a point. Sabelis and Bakker (1992) suggested that such setae are used to extend moving space and to prevent the adhesive threads from sticking to the dorsum. I think that the long dorsal setae of *Tp. bambusae* have an additional function to recognize the prey nest web, as proven in *Stigmaeopsis* species (Sect. 3.5), because this predator has a strong tendency to live inside the prey's nests. Therefore it must have a sensor to recognize "nests" to realize such a habit. Although this is a little bit different

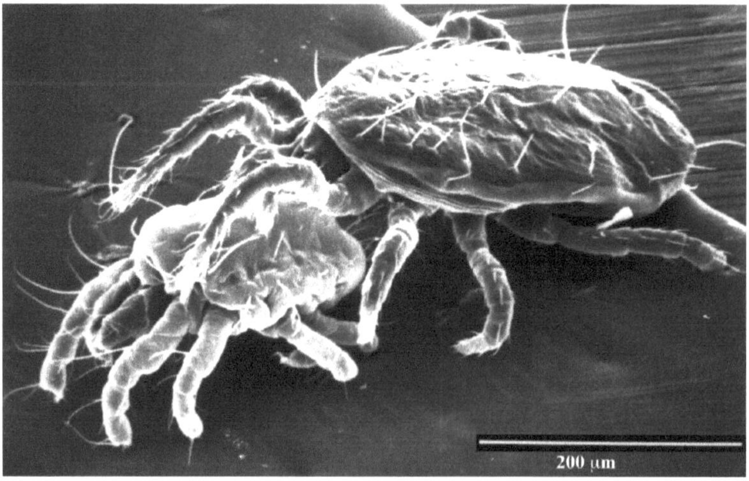

Fig. 2.16 *Typhrodromus (Tp.) bambusae* attacking *Stigmaeopsis (St.) longus* larva. Note that this photograph is a simulated scene (by SEM)

from the "chaetotaxy hypothesis," it is debatable whether *Tp. bambusae* also uses such setae to extend *Stigmaeopsis* nests (note that *Stigmaeopsis* nests are not so sticky) and whether *Pl. persimilis* uses its long dorsal setae as a sensor to recognize a CW web.

How do phytoseiids recognize prey existence? Because phytoseiids have no eyes, they must have some means of recognizing prey colonies. Schmidt (1976) suggested that *Pl. persimilis* (type I) uses CW type webs as a cue of spider mite inhabitation. For such phytoseiids, spider mite webbing is no more than protective refuges, but also an important cue of prey existence. However, how phytoseiid mites can locate their prey from a distance had long been a mystery. Sabelis and Van de Baan (1983) first reported that volatiles from *Tt. urticae*-infested leaves attract phytoseiid mites. Sabelis et al. (1984) observed that starved *Pl. persimilis* females are capable of turning after passing a steep kairomone odor (coming from a spider mite patch) gradient. The chemicals involved in the attraction of *Pl. persimilis* were identified in the headspace of lima bean leaves infested with *Tt. urticae* (Dicke et al. 1990). These results strongly suggested that there are some volatile chemicals that phytoseiids can use as cues to locate prey. Their findings were sensational and surprised many researchers, because such chemicals are not by-products of spider mite injury but are produced by the plants themselves to communicate their predicament to predators (Dicke and Sabelis 1988; Takabayashi and Dicke 1996). It is now apparent that some phytoseiids use chemicals produced by plant leaves to know the general location of spider mite colonies from a distance; they then utilize a tactile stimulus from the webs to pinpoint the exact prey location (Schmidt 1976). These findings showed that there are complex plant–herbivore–predator interactions through tactile and chemical communication.

However, we still have several questions as to how such chemical communication becomes invalid after the extinction of prey on the plant, how *Pl. persimilis* switches its response to such volatiles when it should disperse to other plants where *Tt. urticae* are more abundant, and whether the plants continue to produce such volatiles after the immigration of a sufficient number of predators? In relation to these questions, Maeda and Takabayashi (2001) showed that another phytoseiid species, *N. womersleyi*, uses two cues, i.e., volatile chemicals from other prey-infested plants and prey abundance in the present patch, to decide the timing of emigration. Furthermore, Maeda and Takabayashi (2005) reported that *N. womersleyi* appears to determine residence time in the current patch based on foraging experiences together with the quantity of prey in the current patch.

The aforementioned findings have all been restricted to interactions within mites, but I should add an important discovery in more complex interactions between two predators including insects and their common prey. Magalhães et al. (2004) showed that there is an interesting interaction between thrips (*Frankliniella occidentalis*), a phytoseiid [*Neoseiulus (N.) cucumeris*], and their common predator, *Orius laevigatus* (they called the former as prey and the latter two as interguild prey and predator, respectively). They investigated the behavioral responses of intraguild prey toward their intraguild predator, and which cues elicit these responses, in a system consisting of plant-inhabiting arthropods. Then they showed that *N. cucumeris* can avoid plants with *O. laevigatus* and thrips. This avoidance was triggered by volatile cues associated with the diet of *O. laevigatus*, i.e., consumed thrips. This behavioral response is quite reasonable for *N. cucumeris* to avoid predation by the intraguild predator, *O. laevigatus*. Although they have revealed more complex phenomena in these three species interactions, I stress here that there is another kind of chemical communication even within insect–mite interaction systems. As stated earlier, these findings in chemical communication between prey, predators, and plants strongly show that mites are basically living in chemical rather than visual spheres.

As such, it is known that there is good correspondence between spider mites and phytoseiid morphology and behavior (McMurtry and Croft 1997; Sabelis and Bakker 1992). These results must suggest that at least some Phytoseiidae have coevolved with Tetranychidae in a close predator–prey "arms race." The number of described phytoseiid species in the world, 2,250 (Moraes et al. 2004), is approximately twice that of spider mites (1,189), such that this predator group must also radiate with other prey groups and foods (such as pollen, honeydew and plant juice) (Overmeer 1985; McMurtry and Croft 1997).

Meanwhile, there are three kinds of groups in Phytoseiidae (other than the McMurtry and Croft 1997 categorization), one that has nonfeeding larvae, another that has facultative feeding larvae, and yet another that has obligatory feeding larvae (McMurtry et al. 1970; Zhang and Croft 1994). Nonfeeding larvae in several phytoseiids (*Pl. persimilis*, *N. womersleyi*, *Tp. bambusae*) have drawn some attention. Why some species possess a nonfeeding larval stage has long been a mystery (McMurtry et al. 1970). Zhang and Croft (1994) suggested that females of species with nonfeeding larvae lay larger eggs than females of species with feeding larvae. They also suggested that the mouthparts of nonfeeding larvae could be nonfunctional. However, these hypotheses address the question from physiological and morphological (proximate) rather than behavioral or evolutionary (ultimate) points of view.

On the other hand, Chittenden and Saito (2001) suggested from a sociobiological point of view that such a difference in larval feeding habit must be a kind of social trait to avoid sib-cannibalism. Several studies have shown that cannibalism and interspecific predation are not uncommon phenomena among phytoseiid mites and that immature phytoseiids will readily prey upon younger stages (Schausberger 1999). Because sib-cannibalism is detrimental to an individual's inclusive fitness (Chap. 6), the prevention of such behavior is advantageous if sib-individuals live in an aggregative manner. Chittenden and Saito (2001) showed that species laying their eggs in a mass have nonfeeding larvae and those laying their eggs sparsely (randomly) have feeding larvae. They determined the larval feeding types of ten phytoseiid species and observed their preference to prey life types and egg-depositing patterns (Table 2.4). Four species of ten phytoseiids are native to *Sasa senanensis* (hereafter this plant is simply called "*Sasa*") and the other six were originated from different plants [*Pl. persimilis*, *Neoseiulus* (*Amblyseius*) *californicus*, and *Iphiseius degenerans* were foreign species]. On *Sasa* stands, we know that there are

Table 2.4 Larval feeding type and oviposition pattern in Phytoseiidae (from Chittenden and Saito 2001)

Phytoseiid species	Average no. of eggs/ day (±SD)	Preferred prey life type (%)				Larval feeding type
		LW-j	LW-s	CW-u	WN-c	
Typhlodromus bambusae Ehara	2.00±0.22	0.00	4.17	0.00	95.83	NFL
Neoseiulus californicus (McGregor)	3.38±0.25	0.00	8.64	90.13	1.23	FFL
Phytoseiulus persimilis Athias-Henriot	3.08±0.94	1.33	12.00	78.67	8.00	NFL
Amblyseius sp 1.	3.04±0.50	4.11	0.00	38.36	57.53	NFL
Phytoseius tenuiformis Ehara	1.38±0.22	0.00	0.00	93.94	6.06	NFL
Neoseiulus womersleyi (Schicha)	3.83±0.59	0.00	0.00	71.43	28.57	NFL
Ambliseius orientalis Ehara	1.08±0.19	11.54	73.08	0.00	15.39	OFL
Iphiseius degenerans (Berlese)	1.00±0.25	25.0	29.17	20.83	25.00	FFL
Amblyseius sp. 2	1.25±0.22	20.00	16.67	3.33	60.00	OFL
Amblyseius eharai Amitai & Swirski	1.03±0.41	12.50	50.00	12.50	25.00	OFL

NFL nonfeeding larvae, *FFL* facultative feeding larvae, *OFL* obligatory feeding larvae, *LW* little web type, *CW* complicated web, *WN* web nest. LW-j , subtype appears in *Eurytetranychus japonichus* (Fig. 2.3); LW-s, that in *Yezonychus sapporensis* in Fig. 2.6; CW-u, that in *Tetranychus urticae* in Fig. 2.4; WN-c, that in *Stigmaeopsis celarius* in Fig. 2.6.

various spider mite species having different life types (Sect. 2.3), such that all phytoseiid species targeted could feed, develop, and oviposit by feeding on some of these spider mites. Then we used *Sasa* leaves as experimental arenas and four spider mite species inhabiting *Sasa* stands as prey (species having four life subtypes appear in Figs. 2.3–2.6). We prepared four sections by using water-soaked cotton strips on a leaf area (4.5 cm×4.5 cm). Then we first released several females of each prey species on one of the four areas. Several days after the prey females laid sufficient eggs (and some species made web structures), a female of each predator species was introduced onto the area after eliminating cotton barriers and permitted to select any of the four subareas (thus four prey species).

There were considerable differences in preference between these predator species (see Table 2.4). *Tp. bambusae* only located to the area inhabited by *St. longus*, having a WN life type; *Pl. persimilis*, *N. californicus*, *N. womersleyi*, and *Phytoseius tenuiformis* located to the subarea with *Pn. akitanus*, having a CW life type. These results indicate that there are strong preferences in these phytoseiid mites to particular prey life types. On the other hand, there were three phytoseiid species that located over all subareas, suggesting that they have no preference to prey life types. Simultaneously we checked the larval feeding types of these ten predator species, finding that five species have nonfeeding larvae, three have obligatory feeding larvae, and two have facultative feeding larvae (Table 2.4). To know the egg deposition patterns of these species, the distances between all eggs laid on the experimental arena were measured, such that we could learn there is much variation in egg-laying patterns between species. The species having obligatory feeding larvae tend to lay eggs more sparsely than those having nonfeeding larvae. Species having facultative feeding larvae deposited eggs at an intermediate distance, and those having nonfeeding larvae deposited eggs in masses (Fig. 2.17).

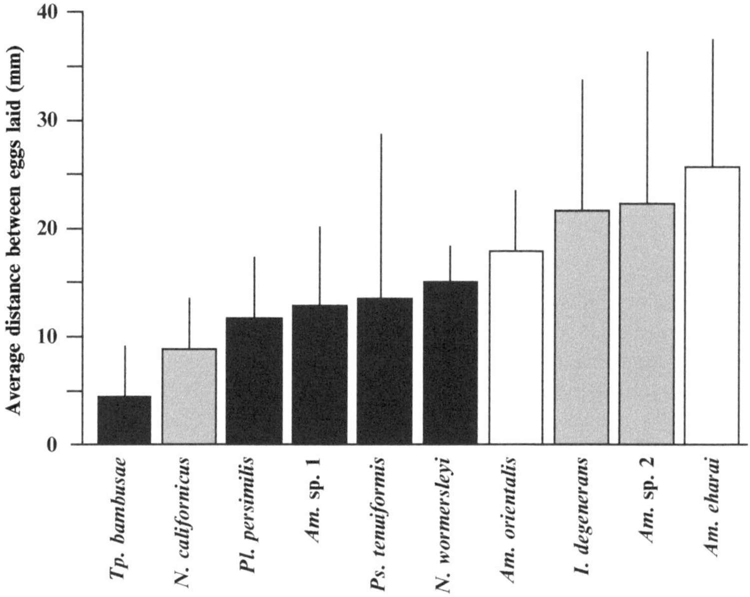

Fig. 2.17 Average distance between eggs of ten phytoseiid species. *White*, obligatory feeding larvae; *gray*, facultative feeding larvae; *black*, nonfeeding larvae. *Vertical bars* are SD. (after Chittenden and Saito 2001)

Depositing eggs in a mass must increase the probability of sib-cannibalism if larvae have high feeding ability (Schausberger 1999). Therefore, the hypothesis that nonfeeding larvae represent an adaptation to avoid sib-cannibalism becomes plausible (Chittenden and Saito 2001). Although such a discussion should remain tentative until a more detailed understanding of the phylogenetic relationships between the species concerned can be acquired (Harvey and Pagel 1991), I believe that nonfeeding and feeding larvae may have evolved repeatedly within the family Phytoseiidae, because nonfeeding larvae simultaneously appear in the genera *Neoseiulus* (*Amblyseius*), *Amblyseius*, *Phytoseiulus*, *Phytoseius*, and *Typhlodromus* (Zhang and Croft 1994). A similar larval nonfeeding phenomenon is known in another predacious mite species, *Hypoaspis (Hp.) aculeifer* (Usher and Davis 1983), and also in spiders (a lack of mouthparts in first-instar larva; Savory 1964), both of which are known as very strong predators, and the latter usually lays eggs in large masses. As such, a kind of convergence among these predacious animals becomes increasingly evident.

In several phytoseiids and *Hp. aculeifer*, we may thus see a beginning stage of social behavior, if avoidance of sib-cannibalism would be considered to be such a phenomenon as supposed by Usher and Davis (1983). More sophisticated behavior is known in *I. degenerans*. Faraji et al. (2000) reported that *I. degenerans* females have a strong tendency to cannibalize conspecific eggs, in discriminately feeding upon unrelated conspecific eggs but not upon its own eggs. In addition, I recently observed that *Tp. bambusae* females never cannibalize their own progeny, even under strong starvation conditions: gravid females were individually confined in glass tubes supplied with water only (Saito and Mori 1975; Chittenden and Saito 2001; see Box 2.10). Here, I observed how long they can survive without food. Several females laid eggs in the tubes, and thereafter larvae hatched. These larvae developed into the protonymphal stage without food, and lived longer than the females without being cannibalized by their mothers (Saito, unpublished data). From such fragmental observations, I now hypothesize that more developed social behaviors will be found in phytoseiids in the future, because there are several examples of social evolution in the same suborder, Gamasida (Treat 1958, 1975; Donzé and Guerin 1994; also see Sect. 4.6).

Turning our attention to an interaction scale other than between individuals within species, there are several fascinating discoveries in phytoseiids. Plant–herbivore–predator interactions other than chemical communications (see foregoing) have attracted attention. Plants sometimes have special physical features, so-called domatia, on their leaves. Such small holes on leaves often serve as refuges for predacious mites and insects (Walter 1996). There was considerable doubt as to whether large plants such as trees developed such physical structures to preserve small arthropods like mites. However, it was recently shown by a mathematical model that leaf domatia probably evolved as antiherbivore devices for trees (Yamamura 2007). Yamamura has explored the conditions for mutualism of host tree, two prey species and one predator, involving apparent competition between the two herbivorous species. Kasai et al. (2005) demonstrated a scenario for prey–predator mutualism in a tritrophic system on a camphor tree (*Cinnamomum camphora*). According to them, *Euseius sojaensis* (a phytoseiid species), which normally feeds on eriophyid mites (sp. 1) living in leaf domatia of camphor tree, effectively regulates another species of eriophyid mites (sp. 2) more injurious to the tree. Therefore, they concluded that there is a mutualistic system of *C. camphora*, eriophyid mite sp. 1, and *E. sojaensis*. Although such a scenario is still hard to prove from a practical perspective, we have several related phenomena in nature. For example, Okabe and Makino (2008) recently demonstrated that the parasitic mite *Ensliniella parasitica*, which uses acarinaria (pocket-like structures on a wasp body), increases the reproductive success of its host wasp *Allodynerus delphinalis* by protecting it from parasitoid wasps; this is also considered to be a tritrophic system among the host, parasite, and predator (parasitoid). Other relevant examples are shown later in relationship to diversity of mites on *Sasa* stands (leaf hairs in Sects. 3.3 and 3.4).

Predators Other Than Phytoseiids

There are certainly many species of predators other than phytoseiids on plants. In comparison with the number of studies done on phytoseiids, relatively few studies have been conducted on other predators. Santos and Laing (1985) reviewed studies on stigmaeid predators (Fig. 2.18) and stated that most species of this group have a

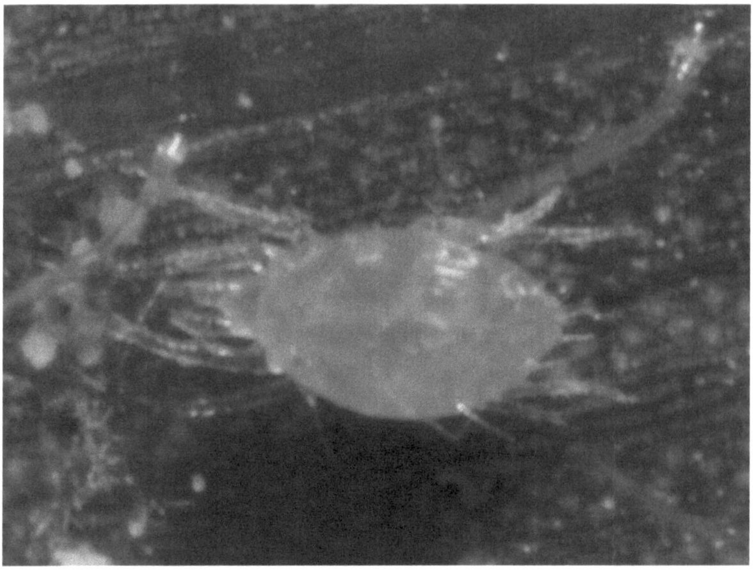

Fig. 2.18 *Agistemus summersi*, a stigmaeid predator on *Sasa senanensis*

low ability to control phytophagous plant mites because of their *K*-selected life histories (low oviposition rates, high longevity, and low dispersal ability). Even though I am not so familiar with stigmaeids, we have discovered that some species play important roles in nature as natural enemies of particular spider mite species as well as other phytophagous mites (Saito et al. 2008a; see Sect. 3.5).

According to Lin and Zhang (2002), Tarsonemidae is a large family of mites present in many different habitats worldwide. Among them, some species occurring on plants are important pests of agricultural crops (Lin and Zhang 2002). We found one predator species, *Tarsonemus praedatorius*, which is the only reported tarsonemid predator of spider mites in Japan (Lin et al. 2002). Females of this species mostly prey on the eggs of *St. longus* in early summer. *St. longus* females cannot recognize such small predators (even though *St. longus* has a high counterattack ability against phytoseiid predators; cf. Sect. 3.5), and they have to face a situation in which no offspring will develop in their nests unless they lay many eggs. I do not yet know what cues they use, but *St. longus* females disperse from such nests to newly grown leaves after heavy infestations by *T. praedatorius* (Saito, unpublished data). Thus, this predator should perhaps be regarded as a parasite of *St. longus* nests, but its importance as a factor enhancing the dispersion of *St. longus* from old to new leaves of *Sasa* should not be underestimated.

Other predators on plants belong to Cheyletidae, Cunaxidae, and Anystidae. I address Cheyletidae in Sect. 4.6 in relationship to social organization. Species of Cunaxidae and Anystidae sometimes occur on plant leaves, but little is known about their ecology other than they are predators. Their nimbleness and ability to prey on phytoseiids and stigmaeids (A.R. Chittenden, personal communication) lead one to suspect that they may belong to a higher trophic level.

Chapter 3
Diverse Microcosmos on *Sasa*

Sasa spp. are typical vegetation components of the forest understory (Fig. 3.1) in Hokkaido, Japan. There are at least three species of *Sasa*, namely, *Sasa senanensis*, *Sasa kurilensis*, and *Sasa nipponica*, and they are distributed separately in response to snow depth in Japan (Suzuki 1978). On leaves of *S. senanensis* (I refer to this species simply as *Sasa*), which is distributed over southern and central parts of Hokkaido where deep snow cover accumulates in winter, at least nine species of spider mites are known to occur (Saito 1990e; Table 3.1).

3.1 Diversity in *Sasa*-Inhabiting Mites

When I began to study the comparative ecology of spider mites about 35 years ago, my target species from *Sasa* was solely *St. longus* (called the long seta form of *Schizotetranychus celarius* at that time). Therefore, I had no idea of the diverse mite fauna occurring on *Sasa*, although I had noticed that there are several different spider mite species there. Hereafter I will describe the mite diversity on *Sasa*, but this does not reflect the order of my study sequence. Many years later, we revealed that the nine species of spider mites are all host specific to *Sasa* and most are sometimes sympatrically found in a single *Sasa* stand. Furthermore, perhaps because of such a diverse and numerous phytophagous mite fauna, there are also more than seven species of predaceous mites living on *Sasa* (see Table 3.1), and five of these are also considered "bamboo-specific" species (Saito 1990e; Chittenden 2002; Gotoh and Shida 2007).

I previously stated that there is diversity in the life history and life type of spider mites (Sects. 2.2 and 2.3; Saito 1983), and I initially thought that such diversity must simply reflect different host plants and environments. However, we have noticed that all the basic life types (see Fig. 2.2) that are currently known in the subfamily Tetranychinae appear on *Sasa*. Furthermore, we have also observed variation in life history among spider mite species on *Sasa* (Fig. 3.2). As mentioned before, life history variation in spider mites has been basically explained by host plant spatiotemporal stability (Saito 1979a; Sabelis 1985). However, it is now

Y. Saito, *Plant Mites and Sociality: Diversity and Evolution*
DOI 10.1007/978-4-431-99456-5_3,

Fig. 3.1 *Sasa senanensis* in Hokkaido

Table 3.1 Mite fauna on *Sasa senanensis* in Hokkaido

Spider mites	Predacious mites
Aponychus corpuzae	*Agistemus iburiensis*
Panonychus bambusicola	*Agistemus summersi*
Panonychus (Sasanychus) akitanus	*Amblyseius ainu*
Schizotetranychus recki	*Amblyseius orientalis*
Stigmaeopsis longus	*Amblyseius* sp. 1
Stigmaeopsis saharai	*Neoseiulus womersleyi*[a]
Stigmaeopsis takahashii	*Phytoseius tenuiformis*
Yezonychus sapporensis	*Typhlodromus bambusae*
Yezonychus sp. 1	*Tarsonemus praedatorius*

[a]Rather rare species on *Sasa*

apparent that such a hypothesis cannot explain the variation in life history in *Sasa*-inhabiting spider mites because they all live on the same host plant. How then can we interpret the diversity in species, life history, and life type (Fig. 3.3) observed on *Sasa*? We started our study with two working hypotheses. (1) The spider mite diversity on *Sasa* was caused by predation pressure. (2) If so, then the predacious mite diversity is a result of various antipredator strategies of the prey in relationship to its webbing behavior, namely, life type and the host plant physical features (for example, leaf hair density and depth of leaf depression).

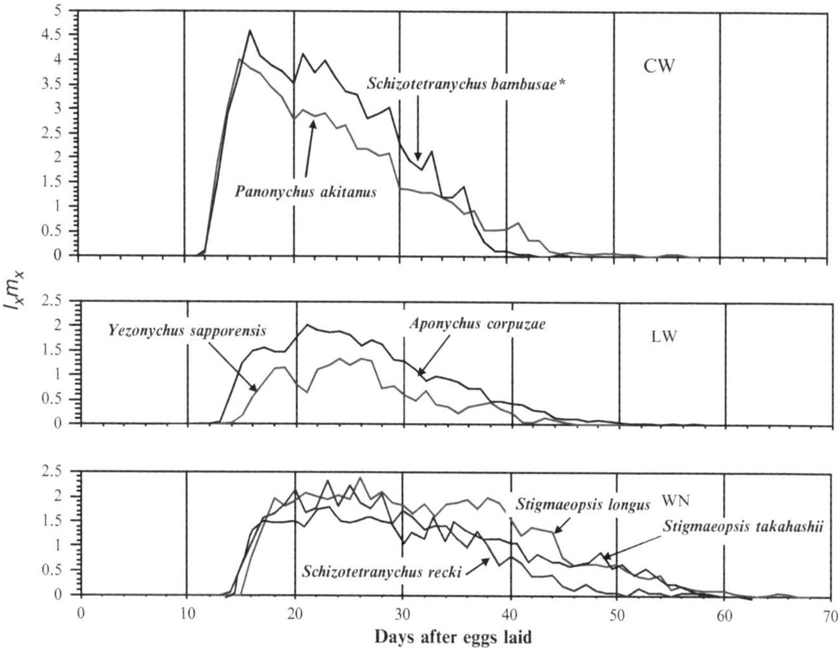

Fig. 3.2 Life history variation in seven species of spider mites on *Sasa* under 25 ± 2°C (Saito and Ueno 1979; Saito and Takahashi 1982; Saito et al., unpublished data). *Asterisk:* species inhabiting moso bamboo (*Phyllostachys pubescens*)

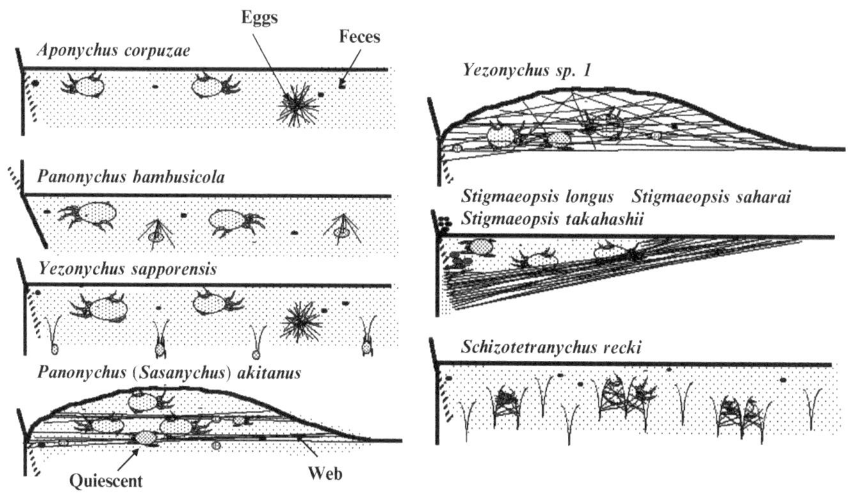

Fig. 3.3 Life subtypes on *Sasa*

3.2 Is the Flattened Body an Antipredator Strategy?

When we observed spider mites inhabiting *Sasa*, *Aponychus (Ap.) corpuzae* immediately attracted our attention because of its special appearance. *Ap. corpuzae* has an extremely flat body (Fig. 3.4). Whether this flat body, and the quiescent and feeding postures, really function as antipredator traits is our question. Furthermore, does the web cover (Sect. 2.3) the female puts over each egg serve any function? Chittenden and Saito (2006) used quiescent (inactive stage) deutonymphs, which normally assume a flat posture on the leaf surface, for this experiment, because it is impossible to keep active mite stages in the optional position. Deutonymphs are quiescent for at least 1.5 days under 25°C conditions, such that we could use them for 24-h predation experiments. Half of them were tipped on their sides using a very fine needle and the other half were left untouched in their natural positions (Fig. 3.5). We also tested the function of egg webs by removing them by a fine needle. Of course, by using a control experiment we had already confirmed there were few effects from such artificial treatments. Then we tested the ability of four predator species that usually co-occur with *Ap. corpuzae* to prey on *Ap. corpuzae* eggs and deutonymphs.

It was apparent that *Amblyseius (Am.) orientalis* frequently ate *Ap. corpuzae*, although it was still difficult for it to eat prey mites in the normal position (Fig. 3.5). We also learned that the egg web cover has a strong function to prevent predation (Fig. 3.6). Two species, *Am. orientalis* and *Tp. bambusae*, could increase their egg consumption rates when the egg webs were removed. Furthermore, the deutonymphs left in the normal position were eaten less frequently by all the predators (Fig. 3.5). Therefore, we reached the following conclusion: The egg webs, the flat body, and the quiescent posture of *Ap. corpuzae* help it avoid at least four predator species (Chittenden and Saito 2006). Thus, we considered that such a body feature evolved as an antipredator trait (Box 3.1).

The fact that most predator species occurring on *Sasa* have difficulty eating *Ap. corpuzae* in its normal position (and with egg cover intact) surprised us. This species has a very wide distribution on various species of bamboo (including *Sasa*); i.e., its distribution is known over North and South-East Asia (it is known to occur

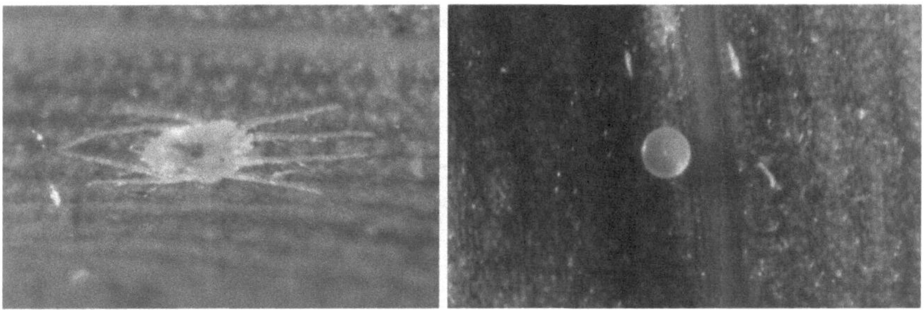

Fig. 3.4 Photographs of *Aponychus corpuzae*: adult female (*left*); egg (*right*)

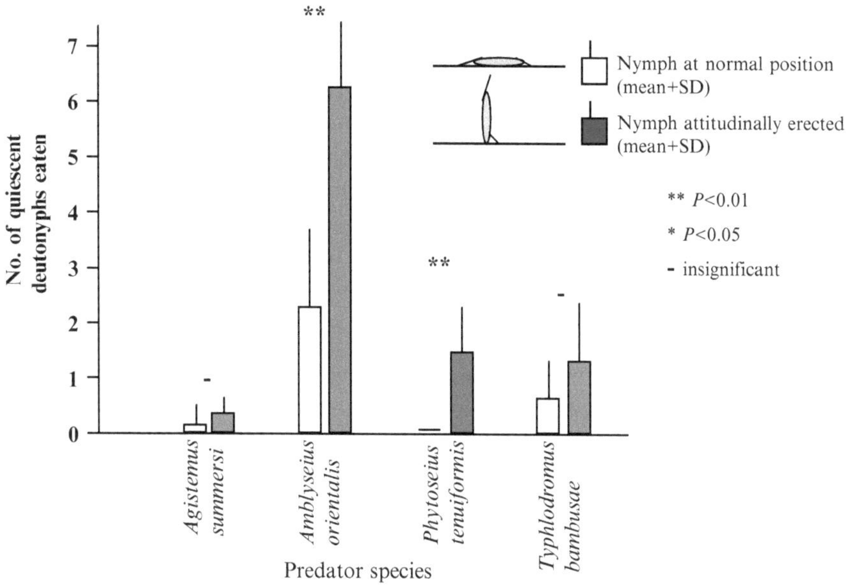

Fig. 3.5 Function of flattened body of *Aponychus corpuzae* as an antipredatory strategy (after Chittenden and Saito 2006)

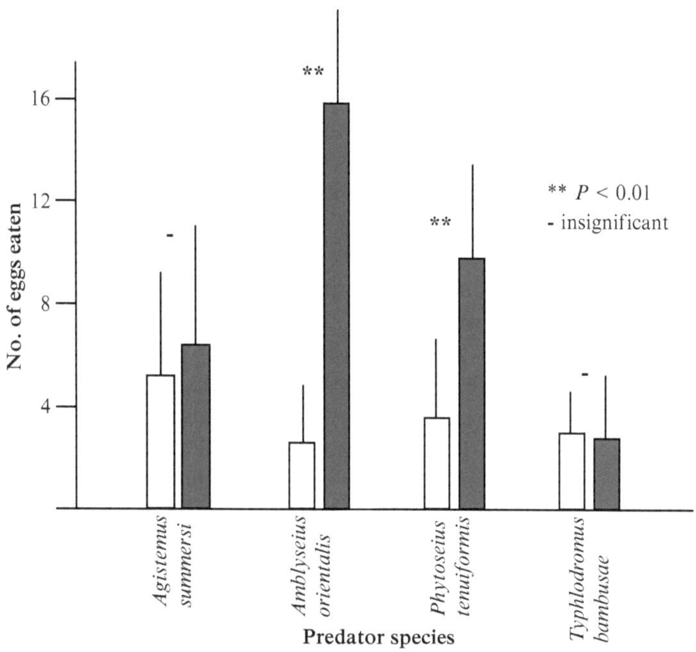

Fig. 3.6 Function of egg web as antipredatory strategy in *Aponychus (Ap.) corpuzae* (after Chittenden and Saito 2006). *White bars*, with egg web cover; *shaded bars*, without web cover

Box 3.1

A flat body shape, like a flat fish, is generally thought to be such an adaptation, and no one doubts this. However, to our surprise, there can be no evidence that a flat body has a "true function" because it is difficult to make such an animal body thicker or thinner. In this context, by using artificial manipulation of *Aponychus corpuzae* quiescent stages, can we find it for the first time?

in Japan, China, Korea, the Philippines, Thailand, Malaysia, and Indonesia, all regions with extensive native *Sasa* and bamboo vegetation). Such a widespread distribution may be explained from its overall ability to escape predation.

3.3 Do Oviposition and Undergoing Quiescence on Tips of Leaf Hairs Constitute Antipredatory Behaviors?

Second, I address the case of *Yezonychus sapporensis*. This species is characterized by where it deposits its eggs and spends its quiescent stages (Fig. 3.7). Larvae and nymphs enter quiescent stages, and females deposit their eggs on the tips of leaf hairs. Why they do so is the next issue.

Yanagida et al. (2001) tested whether eggs on the tips of leaf hairs are any safer than those on the leaf surface. Half of *Y. sapporensis* eggs were manipulated by cutting the supporting leaf hair and lying it flat on the leaf surface, and the other half were left intact, namely, on the tips of leaf hairs. We cut leaf hairs to prevent any artificial damage to the eggs glued tightly to the hairs. We tested seven predator species.

The feeding rates of most predator species decreased when the prey eggs were positioned on tips of leaf hairs. Among the predators, however, *Phytoseius (Ps.) tenuiformis* ate many eggs regardless of their position (Fig. 3.8). Therefore, we believe that the behaviors of *Y. sapporensis*, i.e., depositing eggs and undergoing quiescence on tips of leaf hairs in the summer season, have antipredator functions. Furthermore, only a single predator species on *Sasa*, *Ps. tenuiformis*, can overcome these traits. The first legs of *Y. sapporensis* females are longer than those of other spider mite species (Fig. 3.7), such that we think morphology may also be related to its behavior of depositing eggs on the tips of long leaf hairs.

Meanwhile, when discussing the functions of behavior, we should address the terms "strategy" and "tactic." In evolutionary ecology, strategy generally means a trait that is fixed in a species, whereas a trait that is expressed facultatively in a species is called a tactic (Müller et al. 2006). From these definitions, body shape in *Ap. corpuzae* could be regarded as strategy, but the behaviors seen in *Y. sapporensis* must be considered tactics: in summer, the eggs and immature quiescent stages are

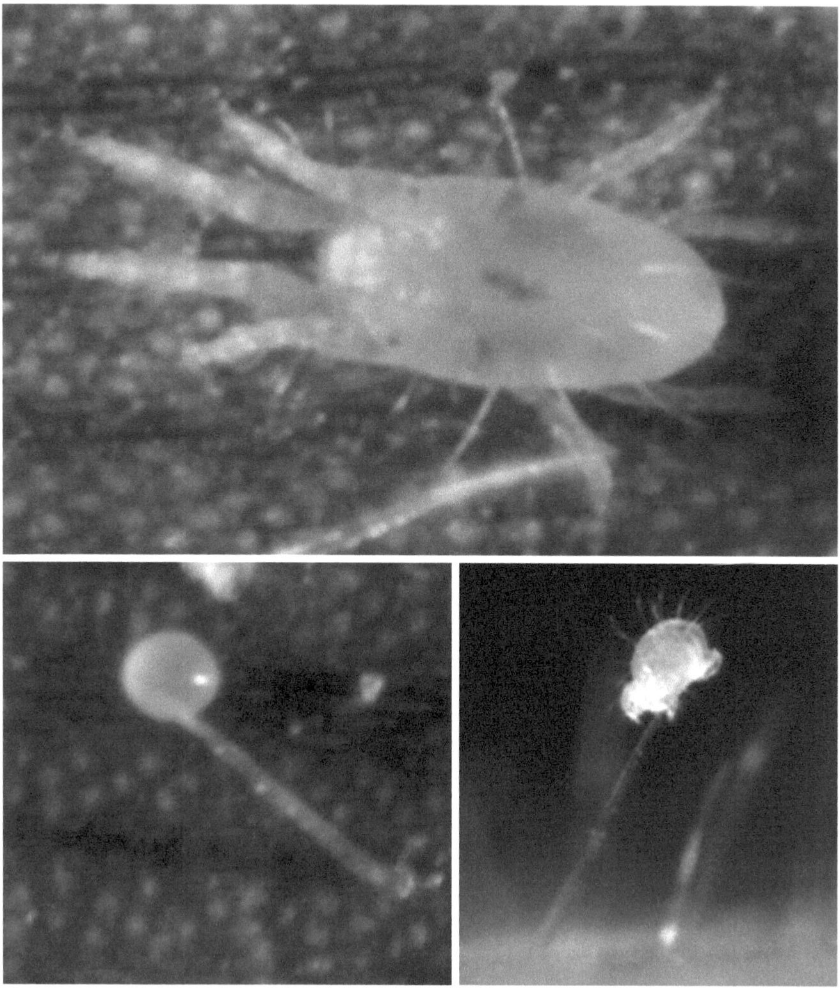

Fig. 3.7 Photographs of *Yezonychus sapporensis* show egg and quiescent larva positioning on the tips of leaf hairs

all positioned at the tips of leaf hairs, but never so in winter (Yanagida et al. 2001). Thus, the individuals of this species alternate their behaviors between seasons, although the ecological reason why they change the behaviors is still unclear. The reason why *Y. sapporensis* females change their oviposition sites from the tips of leaf hairs to the leaf surface in late autumn may be related to diapause induction, because females are known to change oviposition sites from the leaf surface to twigs after diapause inducement in several spider mite species that overwinter in the egg stage (Lees 1953; Gotoh 1989). This change may be analogous to the case of *Y. sapporensis*, although both adults and eggs overwinter in this species (Sect. 2.4).

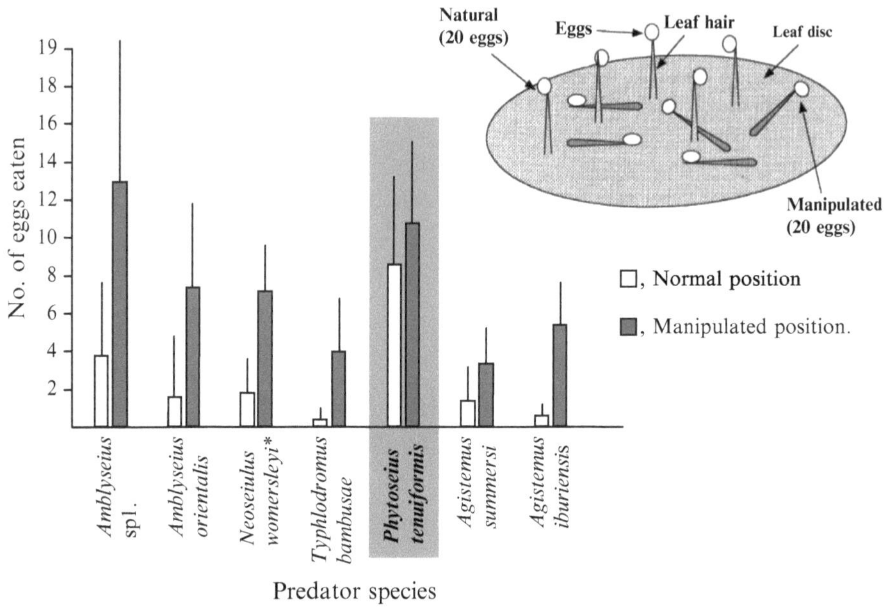

Fig. 3.8 Are oviposition and quiescence on tips of leaf hairs antipredatory behaviors? (Redrawing from Yanagida et al. 2001)

3.4 Do Compartment-Type Nests Have Antipredator Functions?

Third, we address *Schizotetranychus (Sz.) recki* (Fig. 3.9), which makes compartment-type nests by using two or three leaf hairs on *Sasa* leaves. All active individuals make their own nests when they feed and enter quiescent stages. Furthermore, females weave tent-like nests around each egg they produce (Fig. 3.9). Because of such a life habit, it is difficult for females of this species to reproduce on hairless *Sasa* leaves (Saito, unpublished data). Horita et al. (2004) tested whether the nests really have a role in avoiding predation. They artificially removed the nests covering *Sz. recki* eggs and compared the probability of predation between eggs with and without nest protection.

Figure 3.10 clearly shows that most predator species can eat eggs without nest protection more easily than those within nests. The compartment-type nests effectively protect eggs from five species of predators. The only exception observed was *Agistemus (Ag.) summersi*, which tended to eat more eggs contained within nests than eggs outside of nests. Therefore, the most important predator species interacting with *Sz. recki* in nature is considered to be *Ag. summersi* (Horita et al. 2004). Here nature once again shows us two sides of the same coin: Primarily, the antipredator apparatus that evolved in *Sz. recki* is now used by *Ag. summersi* in a contradictory way, i.e., as a cue for prey existence.

So far, I have described several relationships between spider mites and their natural enemies. These results have shown that spider mite diversity as well as predacious mite diversity must be the result of complex evolutionary interactions

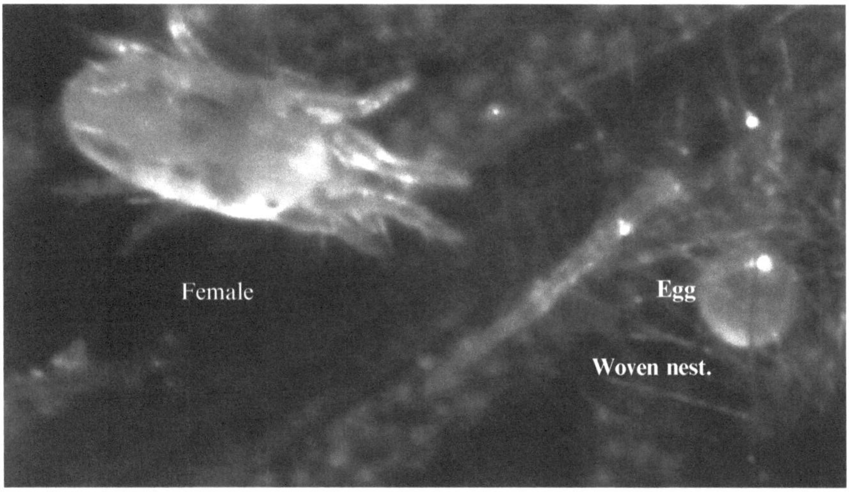

Fig. 3.9 *Schizotetranychus (Sz.) recki* makes compartment-type nests. Female is shown walking outside nest

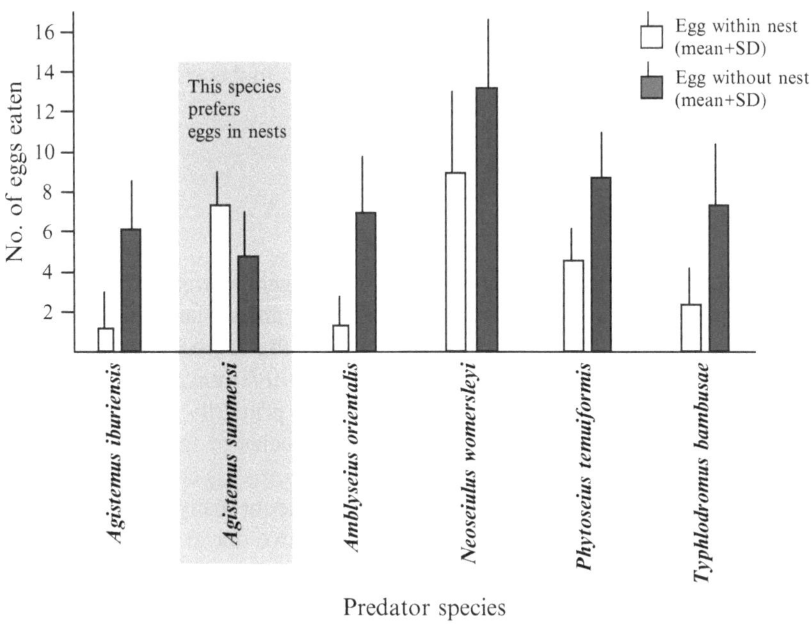

Fig. 3.10 Compartment-type nests of *Sz. recki* function to prevent predators. Choice experiment: between these two treatments all are significant at $P<0.001$ (after Horita et al. 2004)

between them in relationship to the physical features of *Sasa* leaves, i.e., flat surfaces for *Ap. corpuzae* and leaf hairs for *Sz. recki* and *Y. sapporensis*. Dense leaf hairs are necessary for *Sz. recki* and *Y. sapporensis*, as mentioned before, because they are essential materials for the life types of these species, i.e., antipredator traits. In other words, these two species must be limited in their habitat choices by the physical features of *Sasa* leaves (Yanagida et al. 2001; Horita et al. 2004). Actually, there is much variation in leaf hair density among *Sasa* stands, and the distributions of these two spider mites are strongly restricted to *Sasa* stands having dense leaf hairs (Chittenden 2002; Gotoh and Shida 2007). If we focus only on these findings, we cannot understand why some *Sasa* plants grow dense leaf hairs that pest species utilize to protect themselves from predation. However, if we change our focus to other pest species, the situation changes dramatically. We know that the feeding behavior of an aphid species (*Melanaphis bambusae*), which can severely damage *Sasa*, is greatly disturbed by dense *Sasa* leaf hairs (Saito, unpublished data). As mentioned in the previous section, some trees produce leaf domatia that natural enemies use as refuges (Walter 1996). Whether the variation in leaf hair density between *Sasa* stands reflects an antiherbivore apparatus is not easy to determine. All I can say is that the leaf hairs provide important support for herbivorous mites to construct antipredatory refuges.

3.5 Woven Nests of *Stigmaeopsis*

In the previous sections, I have addressed several simple examples of herbivore–predator interactions on *Sasa*. Next, I discuss more complex relationships between *Stigmaeopsis* species and their predators.

What Is Stigmaeopsis?

Before describing the behavior and ecology of the genus *Stigmaeopsis*, I have to provide a brief history of this genus, because it is still unfamiliar even to spider mite researchers. Saito and Takahashi (1982) first noticed that there was variation in nest size and morphology between populations of *Schizotetranychus (Sz.) celarius*. Hereafter, I used, for the time being, "*Sc*" to express primarily a single species, *Sz. celarius*, which was then separated into several species of the "*celarius* species group," until the redefinition of the genus "*Stigmaeopsis*" is completed (Box 3.2). Saito and Takahashi first observed several dorsal setae for many *Sc* populations and found a positive correlation between the lengths of two dorsal setae [P2 (sc1), D2 (d1); Fig. 3.11]. There were at least four distinct groups having different seta lengths in this species distributing over Japan. These findings were the starting point for long-term studies on the morphology, ecology, sociobiology, and speciation of *Stigmaeopsis* (*Schizotetranychus*) species.

At first, we had thought that nest size variation was simply a reflection of complex physical features on the *Sasa* leaf surface: Deep leaf depressions may enable the mite to construct larger nests and shallow depressions may force the mite to build smaller nests. To eliminate the influence of leaf physical features, we developed a method to render leaf depressions constant by using a right-angled acrylic support

Box 3.2

The genus *Stigmaeopsis* had once disappeared from mite taxonomy 54 years ago (McGregor 1950) and was reinstated by Saito et al. (2004). *Stigmaeopsis celarius* Banks was first described by Banks (1917) from an introduced bamboo plant (*Bambusa metake*) as a new genus in Tetranychidae. The Greek name "stigma" refers to the small nest-like structure that is built by this spider mite. Thirty-three years after the first description, this species was redescribed by McGregor (1950) and moved to the genus *Schizotetranychus*. The genus *Stigmaeopsis*, which had only ever contained one species, was then deleted. After that, Ehara (1957) reported *Sz. celarius* occurring on *Sasa* in Japan.

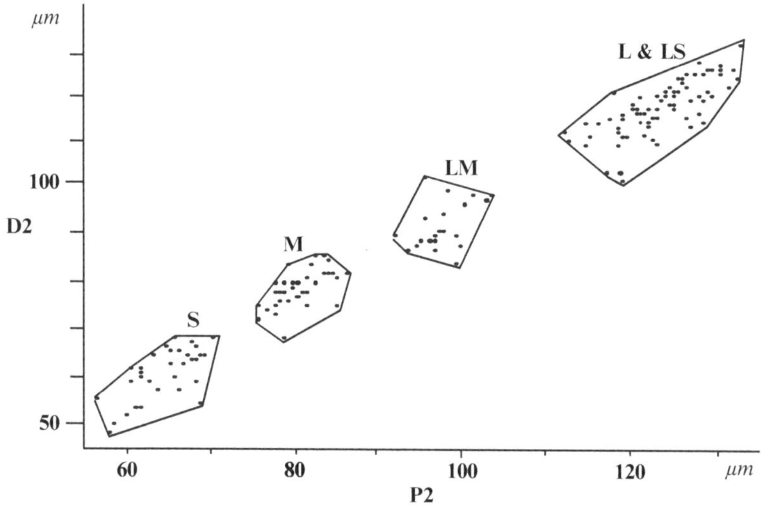

Fig. 3.11 Correlation of two dorsal setae of *Sc* [*Stigmaeopsis (St.) celarius*] in Japan and Taiwan (from Takahashi 1987). *L* long, *LM* long-medium, *M* medium, *S* short (seta forms)

and double-faced adhesive (Saito and Ueno 1979; Mori and Saito 2006; Fig. 3.12). The (shiny) upper surface of a flat *Sasa* leaf was attached to the support with adhesives, to create a right-angled leaf depression on the undersurface of the leaf. An *Sc* female was released onto the right-angled depression of the leaf and permitted to construct a nest (or nests).

Both the nest sizes and the length of the dorsal setae were measured after nest construction, and the measurements were subjected to analysis. We confirmed that the longer are the dorsal setae, the larger are the nests constructed in *Sc* (Fig. 3.13). Hereafter the four groups discriminated by seta length are tentatively called L (long), LM (long-medium), M (medium), and S (short) seta forms of *Sc* in descending order of P2 seta length (Saito and Takahashi 1982) (see Fig. 3.11).

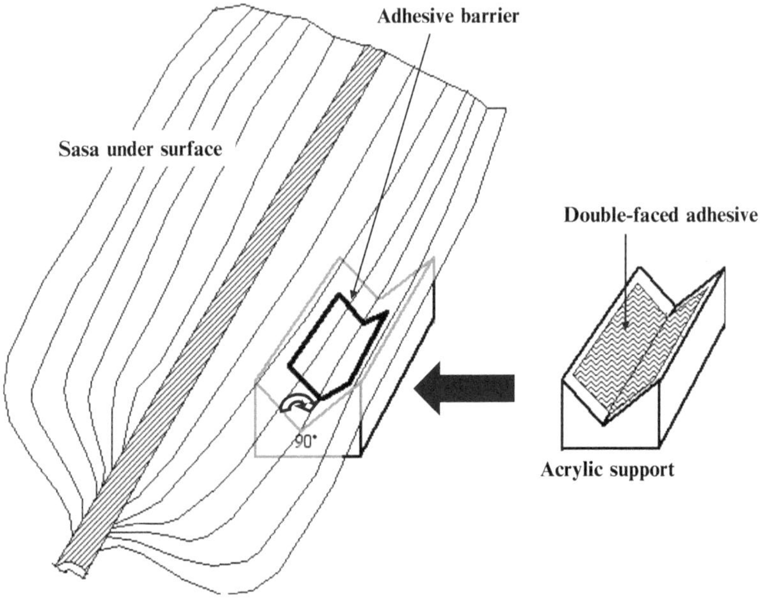

Fig. 3.12 Method to provide constant depressions on leaf

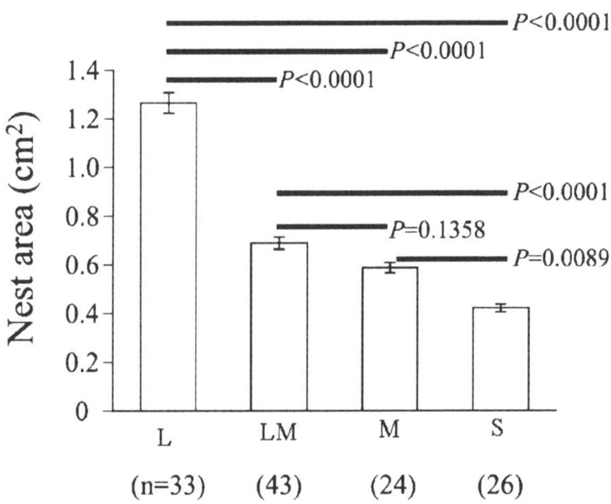

Fig. 3.13 Mean area (±SE) per nest of *Stigmaeopsis*. Results of the *post hoc* test are shown (n = number of females analyzed) (from Mori and Saito 2004, with small changes) *L, Stigmaeopsis (St.) longus; LM, St. celarius; M, St. takahashii; S, St. saharai*

Behavioral observations about mites within their nests showed that several dorsal setae always touch the ceiling of the web nest, leading us to hypothesize that some dorsal setae of *Sc* must function as sensors to recognize nest existence (Box 3.3). Saito and Takahashi (1980) supposed that the dorsal setae extending like the pantographs of electric trains must have such a function (Fig. 3.14).

Box 3.3

As expected, animals living within nests or other closed spaces must recognize whether they really live within such closed spaces. Most animals have eyes, so that there is no need to develop special ways to recognize nests. However, as mentioned in the text, most mites have no functional eyes, such that they must develop other ways to recognize the existence of a nest if they need to live within it.

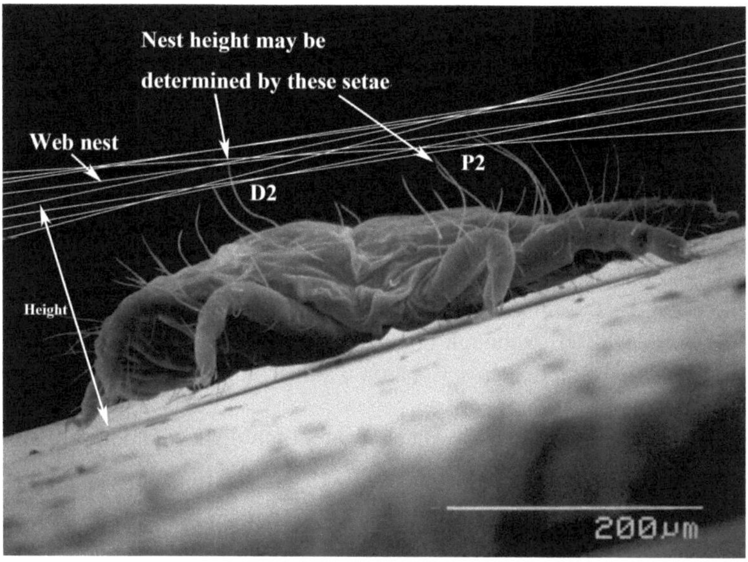

Fig. 3.14 Schematic expression of the function of dorsal setae in recognizing the nest roof. This figure is artificially manipulated with a scanning electron micrograph of *St. longus* and drawn lines

This hypothesis was finally proven by a genetic study between M and S forms. Mori (2000) revealed that the length of dorsal seta, P2, is heritable in an incomplete dominance manner. He discovered that some populations of M (*St. takahashii*) and S (*St. saharai*) produced F_1 hybrids when they were interbred (see later). By use of the hybrid of these two species, he observed the relationship between dorsal seta length and nest size. Then he showed that the seta length of hybrid females became intermediate, and also the nest size built by the hybrid was intermediate between that of the two parental species. This finding clearly showed that the nest size is primarily governed by dorsal seta length, and that such a behavioral trait is heritable.

Thus, we identified the mechanisms that covary with nest size and the length dorsal setae in *Sc*. The next problem was whether the populations showing different

Table 3.2 Reproductive isolation between various forms of *Sc* (*Stigmaeopsis*)

	L	LS	LM	M	S
L	–	**	**	**	**
LS	Sakagami et al. (2009)	–	/	/	/
LM	Mori (2000)	/	–	**	**
M	Saito and Takahashi (1982)	/	Mori (2000)	–	*
S	Mori (2000)	/	Mori (2000)	Mori (2000)	–

–, No isolation; **, completely isolated; *, a small number of hybrid females were obtained; /, not observed

About L, LM, M, S, see Fig. 3.13. LS, *St. miscanthi.*

dorsal setae lengths (and different-sized nests) are considered variations within a single species or different species. Because the differences in setal length are discrete (Fig. 3.11), any reproductive isolation between these forms may determine whether they are separate biological species in the sense of Mayr (1963). Thus, we had to conduct crossing experiments between populations. Table 3.2 shows that there is strong reproductive isolation between almost all combinations of populations having different setal lengths (Saito and Takahashi 1982; Mori 2000). Furthermore, a population collected from *Miscanthus sinensis*, newly designated as LS (dorsal seta lengths are very similar to L) also showed strong reproductive isolation with L (Sakagami et al. 2009). Although one-way reproductive convertibility does exist between M and S, the reproductive isolation is still ecologically important (68–97.9% deleterious; Mori 2000).

From the foregoing results, as well as from morphological and host plant differences, Saito (1990a) could describe two new species, *Schizotetranychus longus* and *Schizotetranychus miscanthi*, from *Sc* (*Sz. celarius*). After that, we had an opportunity to observe two Chinese species considered to be closely related to *Sc*. Based on this, Saito et al. (2004) were able to describe two more new species and defined the genus *Stigmaeopsis* (Fig. 3.15; Appendix 1). Hereafter, I describe these species as those belonging to *Stigmaeopsis*.

Is Nest Size Variation a Reflection of Antipredator Strategy?

Because all *Stigmaeopsis* species can be assumed to be very closely related sibling species, as shown above and by Sakagami et al. (2009), we were able to restrict our comparisons between sibling species to only those phylogenetic constraints that could be minimized. I then tried to solve an evolutionary question as to why there is such variation in nest size between these closely related species.

Nests made from solid and dense webbing undoubtedly function to protect the individuals within them against something. One hypothesis about nest function is that the nests function as protective refuges from predators, as in *Sz. recki* (Sect. 3.4).

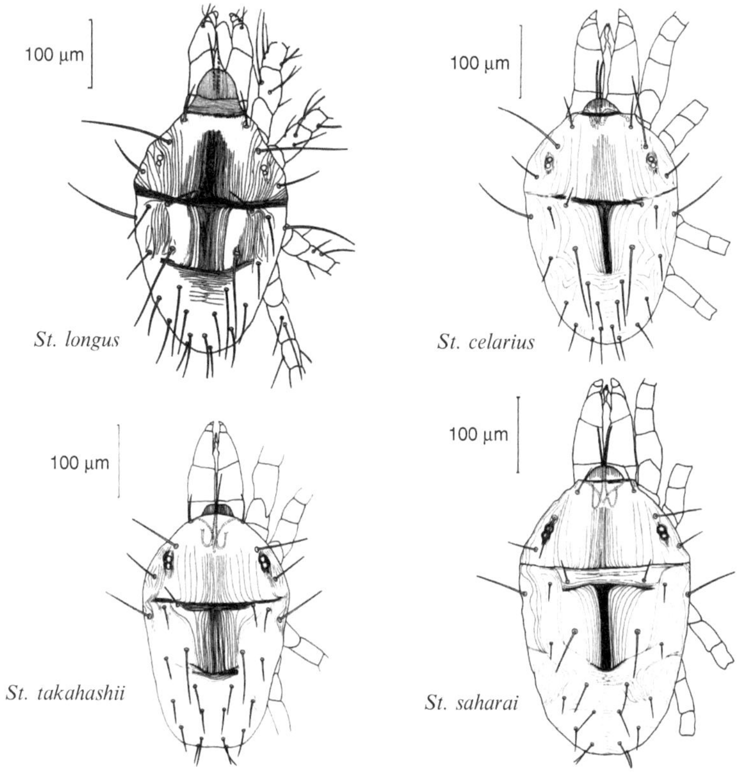

Fig. 3.15 Four species of *Stigmaeopsis* (*St.*) (females) on *Sasa* and bamboo (after Saito et al. 2004). Note the length of dorsal setae

Mori and Saito (2004) could prove this hypothesis by observing that *Stigmaeopsis* spp. nests can more or less protect nest members from certain predator species. Seven to ten females of each *Stigmaeopsis* species were introduced from stock cultures onto detached leaf experimental arenas. After the females had constructed their nests and deposited a sufficient number of eggs within them, the females were removed and two treatments were prepared, i.e., the "web-removal" and "web-intact" treatments, thus somewhat resembling the *Ap. corpuzae* and *Sz. recki* experiments (Sects. 3.2 and 3.4). Adult-stage females of five predator species, which syntopically occur on *Sasa* (see Table 3.1), were introduced individually onto each experimental arena. Two days after predator introduction, we recorded the number of prey eggs eaten. The results clearly showed that the nests have a strong protective effect on the eggs within them (Fig. 3.16). Furthermore, nest size greatly influenced predatory intrusions. The largest *St. longus* nests could only protect eggs from a single predator species, *Amblyseius* sp., but the smaller nests of *St. takahashii* and *St. saharai* could protect against three predator species, although all predator species could eat the eggs of all *Stigmaeopsis* species if there were no web nests.

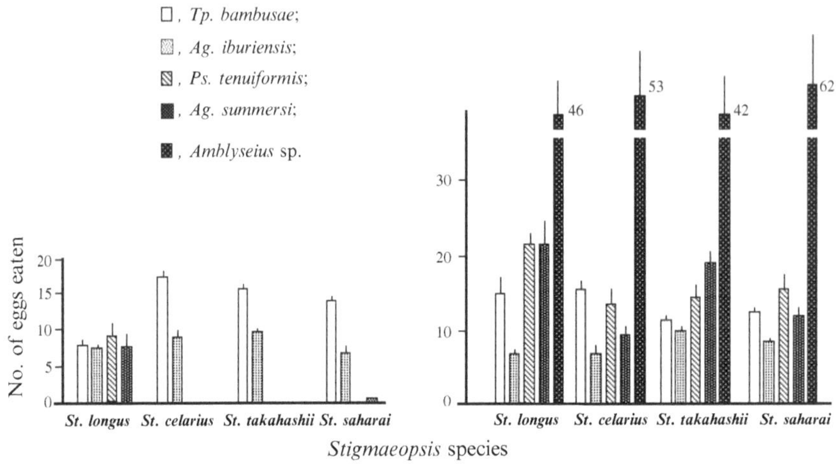

Fig. 3.16 Prey consumption of five predator species over 48 h when there is a nest web (left) or no nest (right) web (from Mori and Saito 2004). Numerals in right side of figure are numbers of eggs eaten. The native host plant of *St. celarius* is considered to be some bamboo species (i.e., *Phyllostachys* spp. native to Japan), and it seldom occurs on *Sasa*. We used this species as an example of a middle-sized nest builder, because it can normally reproduce on *Sasa*

The key point from these results is that the protective effect varies between different-sized nests: Larger nests are vulnerable to the invasions of many predator species, whereas smaller nests can prevent more predator species from invading. If so, from the aspect of the efficiency of the nest as a refuge, the question of why there are species making larger nests inevitably arises. What has happened during evolutionary history – namely, why did nest size diversify? – becomes the next question from an evolutionary point of view.

Advantages of Small Nests

One of the reasons why smaller nests can repel many species of predacious mites is apparently because most predator species can eat prey eggs only when they can intrude into the nest first. *Amblyseius* sp. is a large-bodied predator species that cannot penetrate even the largest nests of *St. longus*. On the other hand, the middle-sized *Ag. summersi* and *Ps. tenuiformis* can intrude into *St. longus* nests through the nest entrances, but are unable to enter *St. celarius*, *St. takahashii*, and *St. saharai* nests. This means that there are no predator species capable of ripping open *Stigmaeopsis* nests on *Sasa*, as *O. perseae* does on avocado (Sect. 2.5; Montserrat et al. 2008). Thus, we could determine that *Agistemus iburiensis* and *Tp. bambusae* must be important common predator species to *Stigmaeopsis* spp.

However, the data shown in Fig. 3.16 are egg death rates over 2 days, such that they are not an accurate indication of the nest effect from the context of their entire lives. If we intend to know the fitness difference between no-nest and nest-intact conditions, it is necessary to convert the egg consumption data into egg survival rates. Let us assume that the survival rate of eggs is zero when there is no nest. When the number of eggs eaten is n (this means that all predators ate the maximum number of eggs), and the number of eggs eaten by predators when the eggs were covered by nests is n_1, $n-n_1$ can be considered the number of eggs surviving as a result of the nest protection effects. Therefore, the survival rate of the eggs covered by nests can be calculated as $sv=(n-n_1)/n$. There is no information about the attack rate of each predator in nature, so I assumed all predators are equally able to attack each *Stigmaeopsis* spp. Under this assumption, sv could be calculated as shown in Fig. 3.17. Although the difference in survival rate between different nest sizes is not so large in this calculation, we may infer that fitness increases with a decrease of nest size in a concave manner (Fig. 3.17). Furthermore, a small increase in survival rate during a short time span (2 days) has a great effect on fitness over a long time span (15 days are necessary for development), if prey mites are repeatedly exposed to the same factors (Box 3.4; Mori 2000). Therefore, smaller nest builders have much greater advantages than larger nest builders.

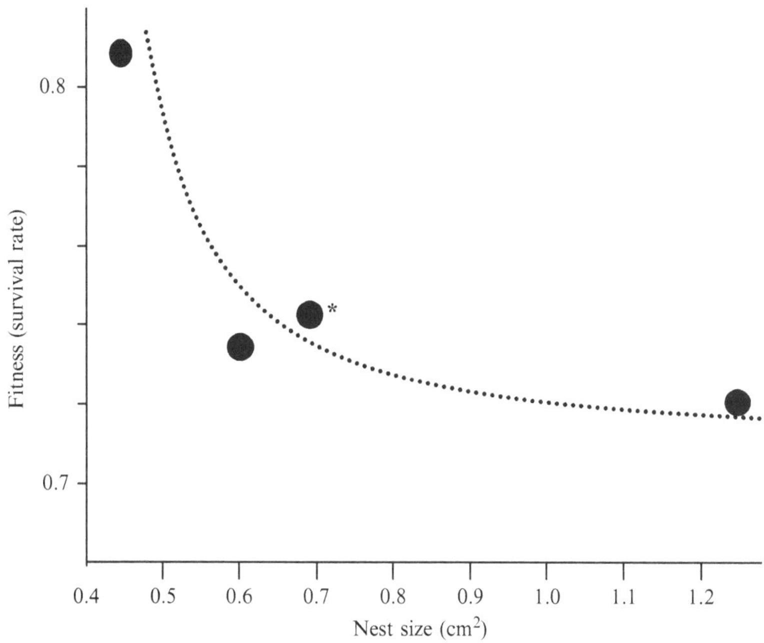

Fig. 3.17 Fitness curve (survival rate) calculated from the prey consumption rates obtained in Fig. 3.16. The procedure to convert the number of eggs eaten to the survival rate of eggs appears in the text. *Asterisk*, *St. celarius*

Box 3.4 Difference in short-term survival rate greatly affects that of long-term survival

Difference in survival rate measured for a short span causes a big difference in fitness for a long period, because the rate is basically a power function of the number of repeats of such spans: Let us denote s_1 and s_2 as survival rates of two species measured during 1 day when they are exposed to the same mortality factors, and t as days when these two species are exposed to the same factors. The survival rate of each species is calculated as $s_1^{\,t}$ or $s_2^{\,t}$. How is such a difference in survival rate crucial? For example, if $t=5$, $s_1=0.5$ and $s_2=0.4$, and the total fecundity = 100; species 1 (s_1) get 3.13 offspring ($0.5^5 \times 100$) but species 2 (s_2) only get 1.02 offspring ($0.4^5 \times 100$). In other words, 0.1 difference in the 1-day survival rate results in a threefold difference in the fitness (number of offspring) after 5 days.

Additional Advantages of Small Nests

Is protection against predator invasion the sole function of smaller nests? The answer may be no. We noticed that there is another kind of variation in *Stigmaeopsis* species nesting patterns. It is closely related to nest size variation, but not always the same. The large nest builders, *St. longus* and *St. miscanthi*, make their nests in a continuous manner, while the smaller nest builders, *St. takahashii* and *St. saharai*, scatter their nests over the leaf surface. As the former two species can scatter their nests and the latter two can build their nests in a concentrated fashon, these differences in nesting patterns are not caused by nest size differences alone. As such, we realized that there are unknown factors forcing species to conduct different nesting patterns. One hypothesis is as follows: for smaller nest builders, nests scattered randomly over the leaf surface (as is actually the case; Saito et al. 2008a) have a hindering effect on predators, even after the nests have suffered predation (note that no predator species can destroy web nests), and this must be related to the searching efficiency of the predator.

To understand the nest-scattering effects, Saito et al. (2008a) conducted several, quite complex experiments to artificially manipulate nesting patterns. If we were to use five *Stigmaeopsis* species, difficulties caused by host plant differences and nest size differences would inevitably arise. Therefore, we decided to use a single species, *St. takahashii*, which builds intermediate-sized nests. Many *St. takahashii* females were introduced into leaf arenas and allowed to build nests and to oviposit. Thereafter, the females were removed, and all the eggs in nests other than a single nest containing many eggs (adjusted to 18 and hereafter referred to as “valid nest”) were crushed to create void nests. Finally, several void nests were arbitrarily removed with a fine brush and distilled water to establish the appropriate number of void nests in the experimental arena. The crushing of several surplus eggs was unavoidable in this procedure. Thus we had to ascertain that the crushed eggs have no effect on the following experiments. One day after the artificial manipulation of nesting and egg-depositing patterns, we introduced *Ag. iburiensis* females (a specific predator, as mentioned before) individually onto each arena and observed their predation for 1 day. If this predator could discriminate nests without eggs (void nests) from those with eggs (valid nests), then there must be no difference in the number of eggs eaten, even if there are various numbers of empty (void) nests.

The results showed that the number of prey eggs eaten decreases significantly with the increase of void nests. This finding showed that it becomes increasingly difficult for this predator species to feed on prey eggs if there are many void nests surrounding each valid nest. Furthermore, an increase of variance among replicates with time was observed when there were many void nests (Fig. 3.18). If predatory success depends more heavily upon chance than on systematic searching, then variance must increase with the number of void nests and with time (conversely, such variance will decrease with time in systematic searching): If there are only a few void nests, predators can easily find the valid nest(s), such that there is little variance among the replicates. On the other hand, if there are many void nests, several fortunate predators may find valid nests quickly, but most will not find any for a certain period. Therefore, in the latter case, the number of eggs consumed will vary strongly between replicates. Moreover, if the predators search for valid nests in a nonsystematic manner (namely, there is no improvement in searching efficiency with time), then variance among the replicates must increase in the latter part of the experiments with many void nests, because the fortunate predators that found a

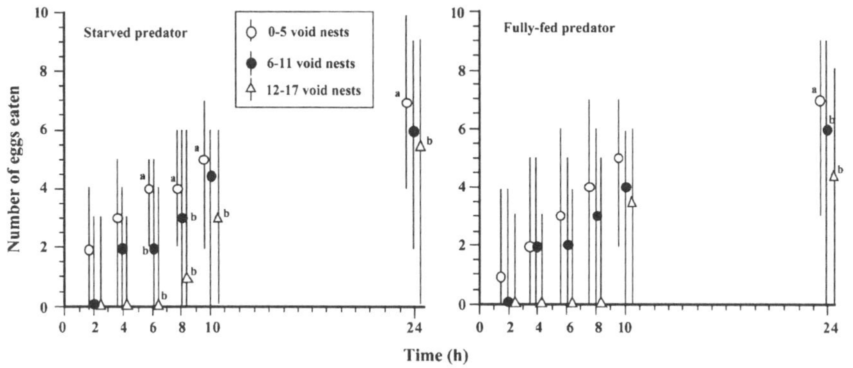

Fig. 3.18 Effects of the number of void nests on prey consumption rate of *Agistemus (A.) iburiensis* for starved (*left*) and fully fed (*right*) predator females. Because the predation processes at the three categories of the number of void nests were unique, median (*symbols*) and range (*bars*) values are shown instead of means and SD. There are significant differences in variance between different characters by *F* test (after transforming the count data by the root) at each observation time ($P<0.01$ with the sequential Bonferroni correction for post hoc multiple comparisons) (from Saito et al. 2008a)

valid nest earlier will have eaten many eggs by that time, whereas predators that had still not found any valid nests will still not have eaten. Therefore, we learned from Fig. 3.18 that nests that have suffered predation, i.e., have become void nests, exert a decoy effect to trick predators.

However, the foregoing experiment could only show the effect of empty nests in *St. takahashii*. Whether the nesting patterns observed in the five *Stigmaeopsis* species really have the effect of decreasing predation pressure was still unknown. To accurately address this problem, we conducted a more realistic experiment using *St. takahashii* to ascertain whether the egg-depositing pattern per se affects predatory efficiency. We arranged the number of nests and the number of eggs within them as shown in Fig. 3.19: i.e., from the number of nests, $x=1$ with 18 eggs to the number of nests, $x=18$, each with one egg. The former is analogous to *St. longus* building only one nest with many eggs and the latter to *St. saharai* building many nests each with small numbers of eggs. Then we introduced an *Ag. iburiensis* female onto each arena. Two days after predator introduction, we observed how many eggs had been eaten. As expected, there was a big difference in predator success between nesting patterns. The predator could consume many eggs when they were concentrated in a single nest, but it could only eat a small number of eggs if they were subdivided into many nests (Fig. 3.20).

The reason why there is such a trend is understandable from the simple probability exercise shown in Box 3.5. If eggs were subdivided into randomly scattered small nests, the predator repeatedly gets trapped inside nests the contents of which have already been consumed (i.e., void nests), and in doing so its chances of finding further valid nests decreases. A simulation result using a practical model of repeated

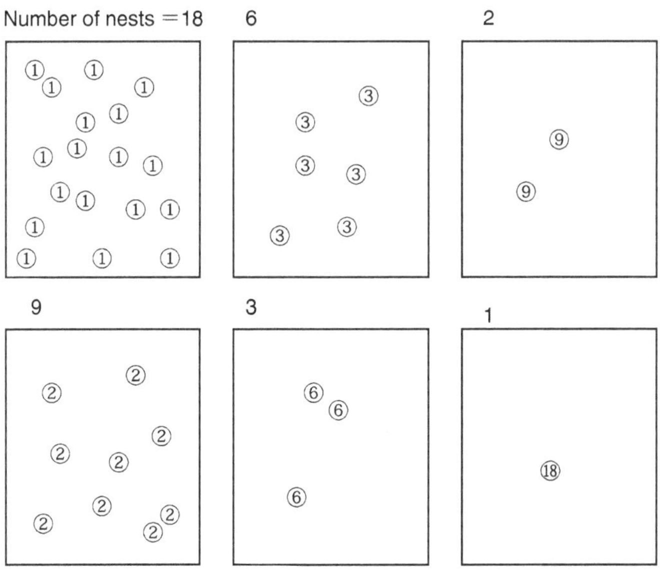

Fig. 3.19 Experimental design to determine the effect of ovipositing and nesting patterns in *Stigmaeopsis*.

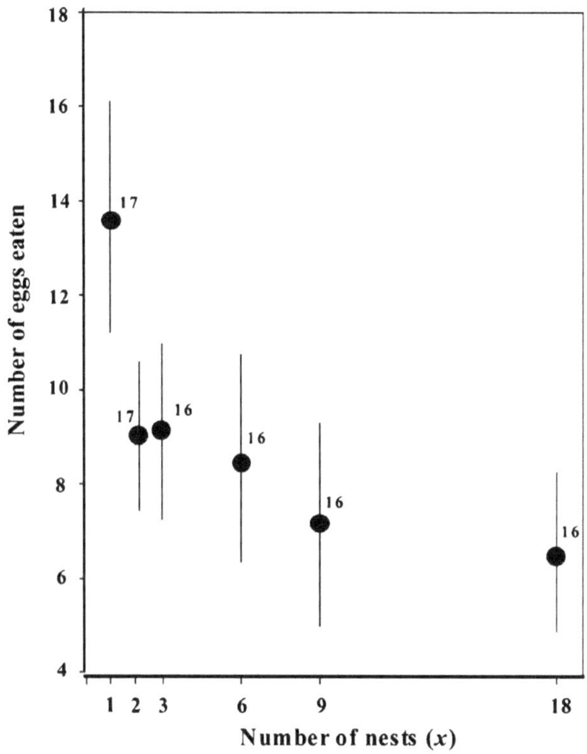

Fig. 3.20 Number of eggs consumed by the predator *Agistemus (Ag.) iburiensis* over 48 h as a function of the number of nests (Saito et al. 2008a). A total of 18 eggs were artificially distributed evenly over all nests (Fig.3.19) for each of the six trials. *Numerals* in the figure show the number of replicates. *Bars* indicate standard deviations of the observed mean. There is a significant correlation between the number of nests and number of prey eggs eaten (by Kendall's rank correlation: $z=-2,24$, $\tau=-0.87$, $P<0.015$)

blind sampling with actual time budgets observed under VTR (Appendix 2) clearly showed the time to eat 18 eggs increases with the number of nests (Fig. 3.21), indicating the effect of nest-scattering behavior.

Therefore, we discovered an overlooked effect of the nesting pattern of small nest builders (Saito et al. 2008a): scattered small nests containing small numbers of eggs have the effect of extending predator searching behavior and thereby protect the offspring contained inside. However, we met a difficulty as to whether we can reconcile the nesting patterns of *Stigmaeopsis* species as a strategy (or function) in the sense of Williams (1966) (Box 3.6). If a female makes many empty nests and scatters them before commencing oviposition, this can be considered a defensive trick against predators. However, small nest builders of *Stigmaeopsis* species do not adopt such a "strategy." Our discovery is that all females make only valid nests (i.e., nests with eggs), and as a result of predation these change into void nests.

Box 3.5

Repeated build sampling trials: At the beginning there are only white balls in the bag. At each sampling trial, when a player samples a white ball, he exchanges the ball with a black one and returns the black ball into the bag. If the player samples a black ball then he simply returns it into the bag. Subsequently, the probability of drawing a white ball decreases rapidly with the number of sampling trials. If we substitute the white balls with valid nests and the black ones with nests void of eggs, this process is an exact replica of the tricking effect modeled in Appendix 1 and shown in the text.

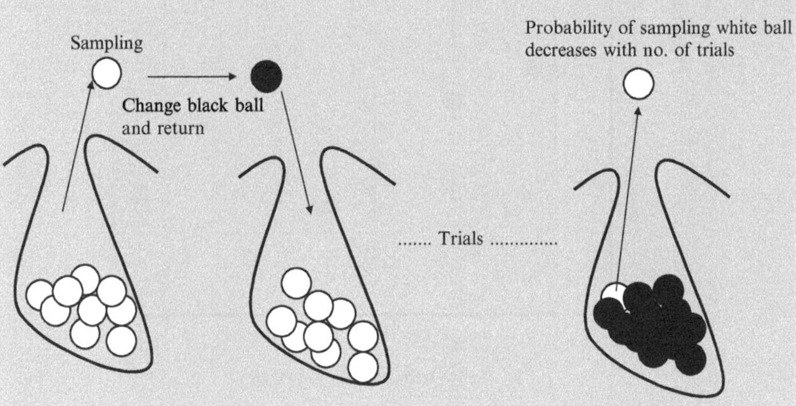

Therefore, someone may consider that it is not a strategy but only a by-product of ordinary nesting behavior; this means that we should strictly consider this effect as an indirect effect of nesting pattern. However, I now believe that such nest-scattering behavior per se could evolve as a tricking strategy by natural selection because individuals scattering their nests in such a manner undoubtedly get higher offspring fitness. Whether such an effect is authorized as one of a new suite of predator avoidance adaptations such as mimicry, crypsis, schooling, effective escape (Taylor 1977; Schmidt 1990), and the laying of egg masses (selfish-herd in Hamilton 1971; delusion effect in Calvert et al. 1979; and Foster and Treherne 1981; Vulinec 1990) should be still tentative.

In any event, the nesting behavior of constructing small nests has another advantage: they are more easily scattered than large nests. Therefore, we know that there are double advantages for small nest builders. We again infer that fitness increases with a decrease of nest size in a concave manner (by converting the small number of nests to large nests, and vice versa in Fig. 3.20). If so, then the crucial question now becomes why there are large nest builders at all in *Stigmaeopsis*.

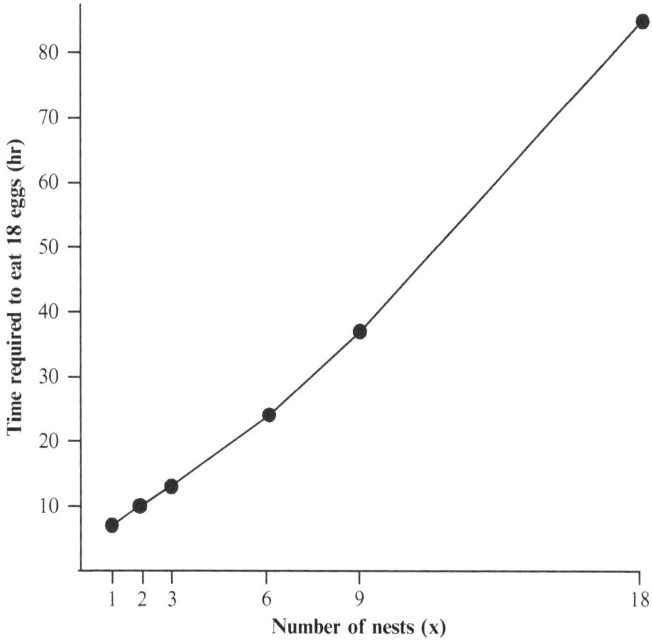

Fig. 3.21 Simulation of hindering effect of nesting pattern on time required to consume 18 prey eggs in *Ag. iburiensis*. *x* is number of nests per arena. Number of prey eggs in a leaf arena is 18 (constant), such that number of eggs per nest varies from 1 to 18 depending upon *x*. Searching and eating time only were considered (Table A-1 in Appendix 2)

Box 3.6

Williams (1966) said, "Whenever I believe that an effect is produced as the function of an adaptation perfected by natural selection to serve that function, I will use terms appropriate to human artifice and conscious design. The designation of something as the *means of mechanism* for a certain *goal or function or purpose* will imply that the machinery involved was fashioned by selection for the goal attributed to it. When I do not believe that such a relationship exists I will avoid such terms and use words appropriate to fortuitous relationships such as *cause and effect*."

Why Are There Large Nest Builders?

The answer to this question may come from the discovery of sociality in the large nest builders *St. longus* and *St. miscanthi* (Saito 1986a, 1990c). *St. longus* is known to show strong counterattack behavior against predators, and attack efficiency increases with the number of defenders (Fig. 3.22; Saito 1986b). Because a united

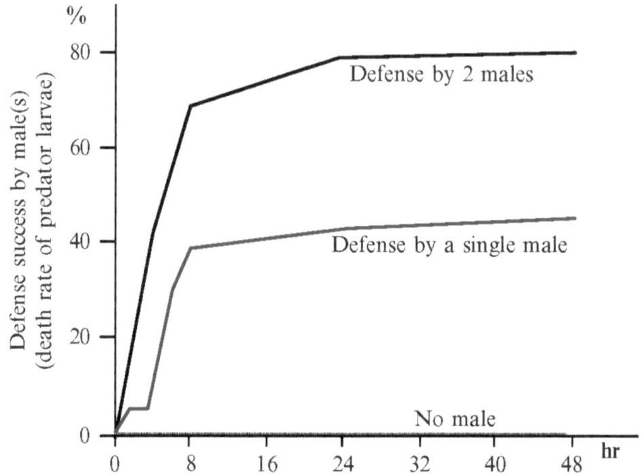

Fig. 3.22 Male density and counterattack success (from Saito 1986b) shown by 2-day survival of predator larvae. There can be no further increase in predator mortality because after 48 h all larvae have developed to protonymphs, which show strong predatory activity

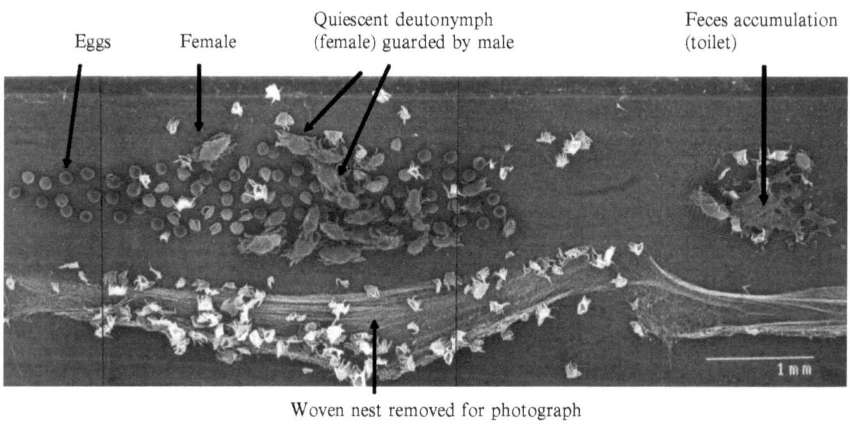

Fig. 3.23 Typical WN-c life type of *Stigmaeopsis (St.) longus* inhabiting *Sasa* bamboo leaf (SEM micrograph). This photograph shows only approximately one-fourth of the united nest

St. longus nest can sometimes exceed 20 cm in length, large numbers of adults must be required to defend their nests effectively against predator invasion (note that this species has a 0.5-mm body length and no functional eyes). Therefore, it is easy to understand that there is a great advantage for *St. longus* to live gregariously within its large united nests (Fig. 3.23).

In relation to this, Mori and Saito (2004) compared the counterattack efficiency of each *Stigmaeopsis* species against the larvae of a specific predator (*Tp. bambusae*). Figure 3.24 indicates that *St. longus* is the strongest defender among the Japanese

Stigmaeopsis spp. and that the defensive effect decreases with nest size. It was also known that two large nest builders, *St. miscanthi* (HG form; see later) inhabiting Chinese silvergrass (*Miscanthus sinensis*) and *St. nanjingensis* inhabiting moso bamboo (*Phyllostachys pubescens*) in China, also show very strong counterattack behavior against phytoseiid predators. In the former, a single male had a 0.67 probability of excluding a predator larva for 48 h (Box 3.7) and 0.75 probability for 24 h in the latter (Zhang 2002). Therefore, there is no doubt that all the large nest builders have strong counterattack abilities against invading predators. As described later, this behavior is one of the important criteria for defining these species as cooperatively social (Sect. 4.4). The middle-sized nest builder *St. celarius* (note that this species is not native to *Sasa*) also shows counterattack behavior against predators, although its success rate is lower than that of *St. longus* (Fig. 3.24; Mori and Saito 2004). On the other hand, the two builders of small nests, *St. takahashii* and *St. saharai*, showed no such effective antipredator behavior. Both males and females tended to move out of their nests when predator larvae had invaded. Thus we determined that there is apparent variation in social behavior among *Stigmaeopsis* species (Mori and Saito 2004).

Box 3.7

Counterattack success of *Stigmaeopsis miscanthi* against *Typhlodromus bambusae* larvae (Yano and Saito, unpublished data). *St. miscanthi* (HG form, Tobuko population) and *Tp. bambusae* larvae were used. Nest survival rate = probability of predators killed (*black bar*) or driven out (*white bar*) from nests for 48 h (25 replicates for each prey combination).

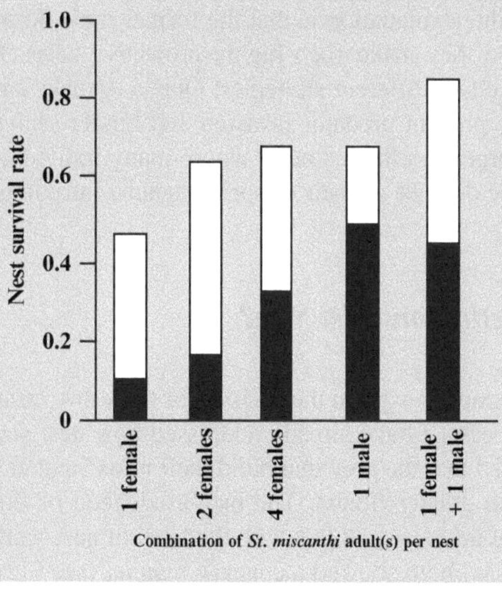

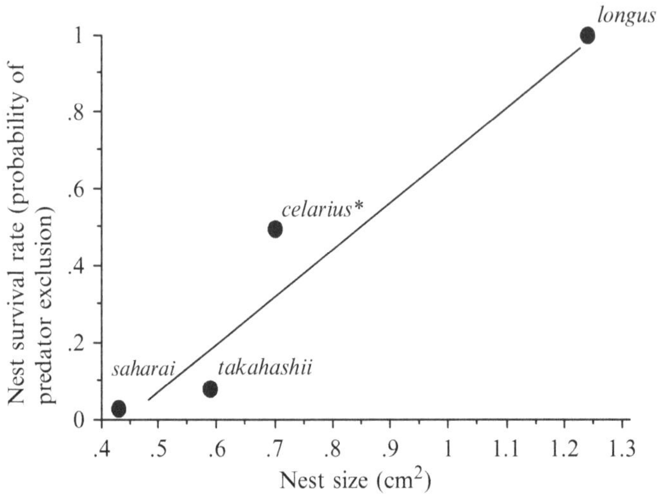

Fig. 3.24 Counterattack efficiency (=nest survival rate) varies among *Stigmaeopsis* species having different nest sizes. Nest survival rate = killing rate + driving out rate of predator when two males of each species defend a nest during 1 day (after Mori and Saito 2004). Remember that the native host plant of *St. celarius* is considered to be some bamboo species and it seldom occurs on *Sasa*. We used this species as an example of a middle-sized nest builder because it can normally reproduce on *Sasa*

Why is there such variation in the counterattack behavior? One answer may be that the large nest builders preserve their inclusive fitness through counterattack (see Chap. 6) because such large nests can contain many offspring. Conversely, small nest builders should escape from predators, because they keep only a few offspring in their nests and will have ample opportunity to reproduce again somewhere else. Another possible explanation is that the former must do so, but the latter has no need to, because they make such highly protective nests. In other words, we expect that there are two different strategies: One is to make small, scattered nests that can effectively prevent predator invasion and hinder searching behavior. The other is to make large, continuous nests where many individuals can cohabit and perform cooperative defense of their offspring against intruding predators.

Disruptive Selection on Nest Size?

Hereafter I will attempt to explain the division of these two strategies in relation to their speciation process. I have already addressed how nest size affects predatory intrusion and showed that the prey that build their nests smaller and in a more scattered manner obtain greater fitness. The increased trend of fitness (survival rate) with the decrease of nest size and that with the level of nest scattering (linking with the nest size decrease) both showed a concave manner (see Figs. 3.17, 3.20).

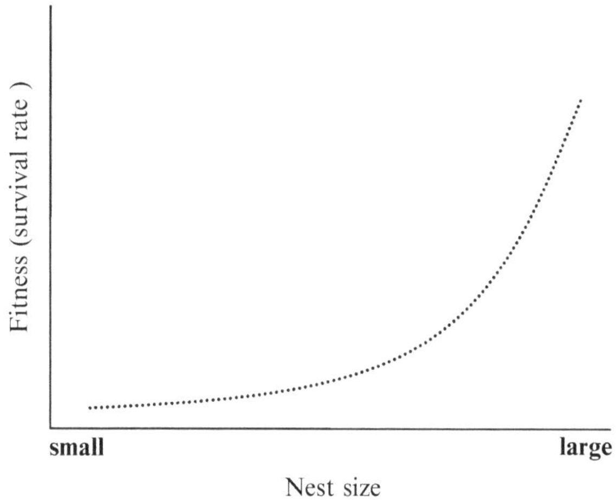

Fig. 3.25 Expected effect of nest size on survival rate of *Stigmaeopsis* species by counterattack (when the length of parents attendant and number of defenders in nests is considered) (Mori 2000)

Conversely, by focusing on the effect of counterattack, fitness may increase with nest size in a concave manner, as shown in Fig. 3.25, although Fig. 3.24 showed a linear increase trend: The parents of large nest builders tend to attend their offspring during their entire developmental time, but those of small nest builders only attend their offspring at most during their egg period (reminder: small nest builders repeatedly make nests in a scattered fashion). It was discussed previously that the difference in short-term survival rate is always an underestimation if we consider the length of interaction (see Box 3.4). The effects of counterattack in Fig. 3.24 were only observed for 24 h under experimental conditions. Thus, the differences in survival rate by counterattack between large nest builders and small nest builders must be much larger if there are constant predator invasion events. Furthermore, the effect of the number of defenders, which is usually very large in large nest builders (cf. Fig. 4.7), was not considered in Fig. 3.24. These unconsidered factors must increase greatly the effect of counterattack in the large nest builder *St. longus*. Thus, it was strongly suggested that the effect of counterattack of large nest builders is much larger than that observed for a short period with a small number of parents (note again that *St. celarius* is not native to *Sasa*).

Under the foregoing reasoning, we could get two contradictory concave fitness curves against nest size (Fig. 3.26). Because the effect of either one of these curves is enough to save offspring, the total fitness of these effects is additive. Thus, total fitness must be a quadratic curve as in Fig. 3.26 (Mori 2000). On such a fitness curve, the smallest and largest nest builders may achieve higher fitness than intermediate nest builders. Mori (2000) thus believes that this is the reason why we can simultaneously observe two extreme nest sizes on *Sasa* leaves, namely, large and small nest builders.

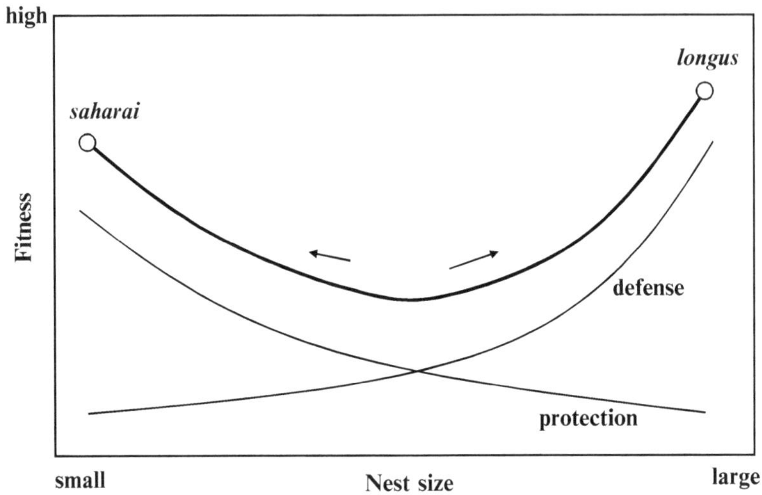

Fig. 3.26 Hypothetical relationship between offspring fitness and nest size variation (from Mori 2000)

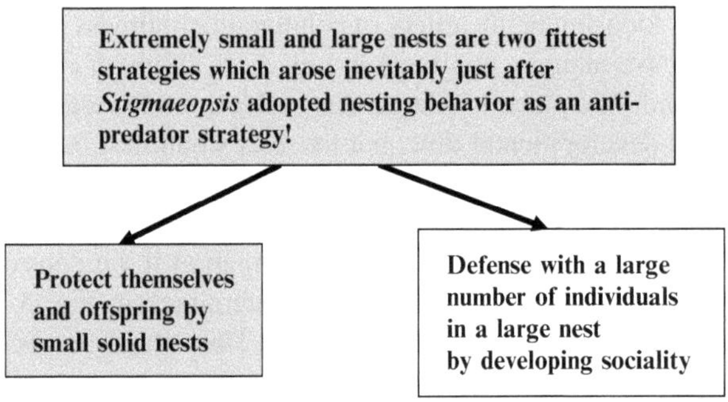

Fig. 3.27 Expected process of speciation through nest size effects

As such, extremely small and extremely large nests are thought to be the two fittest strategies that inevitably evolved soon after *Stigmaeopsis* adopted nesting behavior as an antipredator strategy. Such traits, namely large nests with cooperative sociality and small, scattered nests functioning to prevent predatory intrusion, might lead these mites in different directions. Therefore, Mori and Saito (2004, 2005) proposed the possibility of predator-mediated speciation. Figure 3.27 shows the outline of this process. If there was an ancient species having middle-sized nests and a mutation for making smaller nests occurred, the builder of smaller nests would benefit from lowered predation pressure. On the other hand, if another mutation to make larger nests occurred that incorporated defensive behavior, such a mutation also benefits from lowered predation pressure. Therefore, two extreme strategies, namely, making both the largest and the smallest nests, might be selected if all other things are equal. This scenario suggests that some *Stigmaeopsis* species may have evolved through "predator-mediated speciation" (Mori 2000; Mori and Saito 2005). The discovery that nest size is determined by incomplete dominance gene(s) that govern the length of dorsal setae (as sensors), as mentioned previously, lends indirect support to this hypothesis because such traits are heritable.

A Further Question

As such, we can explain why two species having very different-sized nests occur sympatrically on *Sasa* leaves. Actually, we could regularly observe two *Stigmaeopsis* species, either *St. longus* (large nest builder) and *St. takahashii* (small nest builder) or *St. longus* and *St. saharai* (small nest builder), in *Sasa* stands sympatrically, but *St. takahashii* was seldom found sympatrically with *St. saharai* (Chittenden 2002). The largest extreme is observed in *St. longus*, the females of which each build large and continuous (united) nests over their lifetimes. A further question then arises: What is the minimum nest size in *Stigmaeopsis*?

As described before, there are two small nest builders, *St. takahashii* and *St. saharai*, occurring on *Sasa* leaves, even though they are distributed allopatrically (Chittenden 2002). There was a significant difference in nest size between these two species, i.e., the nest of *St. takahashii* is larger than the nest of *St. saharai* (see Fig. 3.13). The reasons for this difference must be related to minimum nest size. First, we can easily conclude that the minimum nest size must be larger than mite body size, because if the nest is the same size as an individual body, then it is not a "nest" but more like a body wrapping (like a suit). As previously described, small nests prevent predators from invading, but this is only effective when the nest has a protective refuge function. A nest that is too small has no such effect and no protective function against predators, because predators can feed on mites directly through the webbing. Such a phenomenon has occasionally been observed in interactions between *St. saharai* and *Ag. summersi*: the latter is a relatively large predator, occurring on *Sasa*, that sometimes preys upon the former from outside the nest (see Fig. 3.16), although it seldom preys upon *St. takahashii* in the same manner (Yoshida et al., unpublished data). This observation suggests that the minimum nest size might be determined by the risk of predation from outside the nest. The fact that *St. saharai* is never distributed with *Ag. summersi* on *Sasa* leaves in the field (Chittenden 2002) suggests that this proposition may have reality. Furthermore, small nest builders must frequently go outside their nests to make new nests, but walking outside is risky for females because many predacious mites occur on *Sasa*.

3.6 Explanation of Diversity in Life Type and Life History Variation on *Sasa*

In the previous sections, I have addressed several examples thought to have resulted from ecological interactions between predator and prey species. If so, is it plausible to consider the notion that such interactions may be the sole cause of mite diversity? Are there any other explanations for this phenomenon? As a reply, let us again address the diversities in life history and behavior, as well as some other special features of these mite species.

We have observed variation in life history and life type among spider mite species on *Sasa* (see Table 3.3; see Figs. 3.2, 3.3). Furthermore, I have shown that there are differences in the predator–prey relationships among *Sasa*-inhabiting species. The variation in life history between these spider mites cannot be explained from the host plant stability hypothesis (Sect. 2.2; Saito and Ueno 1979), because all the mites live on the same host plant. On the other hand, there does appear to be a trend that species having high r_m show CW life types and those having lower r_m show WN life types (see Fig. 3.2), whereas species having LW life types show intermediate r_m values. Therefore, these trends may be closely related to the differences in anti-predatory traits: As stated, mites building CW type webs have no special behavior against predators, and such webbing is merely the result of mites walking and/or aggregating (Sect. 2.3). If the web has some additional roles in mite living, such as the location of eggs and feces deposition or the exclusion of some predator species, then the production of silken threads is itself not costly; i.e., the web effects are basically by-products of lifeline threads, and as such may not be subject to the cost of "time." Therefore, I expect that these mites are colonizers of predator-free space on *Sasa*. They increase their populations rapidly on newly occupied *Sasa* leaves where predators are scarce and then disperse after predatory mites begin to accumulate. This function is considered to be a kind of *r*-strategy, if we can replace "high probability of predator accumulation" with "easily exploitable and ephemeral habitat" in the original *r/K* theory.

On the other hand, if mites could develop some apparatus and/or behavior(s) to decrease predation pressure, they can share the same habitats (*Sasa* leaves) for

Table 3.3 Life type and overwintering peculiarities of spider mites on *Sasa*

Species	Leaf surface infested	Life type	Overwintering stage
Aponychus corpuzae	Underside	LW	Female
Panonychus bambusicola	Upperside	LW	Egg
Panonychus akitanus	Underside	CW	Female and egg
Schizotetranychus recki	Underside	WN	Egg
Stigmaeopsis longus	Underside	WN	Female
Stigmaeopsis saharai	Underside	WN	Female
Stigmaeopsis takahashii	Underside	WN	Female
Yezonychus sapporensis	Underside	LW	Female and egg
Yezonychus sp. 1	Underside	CW	Unknown

longer without dispersing away. Mites showing WN life types have evolved such a trait. Because constructing solid nests requires both time and resources, there must be a trade-off (in the evolutionary sense) between the ability to reproduce rapidly and web construction. Furthermore, we know that most species having WN life types are relatively small in size, possibly because they live in nests. Such morphological traits may also decrease their fertility. Therefore, they evolved as *K*-strategists by acquiring a good ability to resist high predation pressure.

In relationship to the foregoing, I reported that there is an interesting correspondence between spider mite life history and predator life history (Saito 1990b): The life history of *Tp. bambusae*, which feeds upon *K*-selected *St. longus*, closely resembles that of its prey, i.e., low r_m, but *N. womersleyi*, an important predator of *Tt. urticae* (considered an *r*-selected species), has a very high r_m compared to that of its prey. Of course, these are only two prey species and their specific predators, but such a phenomenon can provide a valuable insight into understanding mite life history evolution (also see Sabelis and Janssen 1993).

Generally speaking, when reasoning about *r*/*K* selection on *Sasa*, another kind of evolutionary trade-off must be considered: rapid increase with high dispersal costs to escape predators may favor CW types. On the other hand, parental defense and cooperation without dispersal must incur considerable costs to realize the WN type. LW type mites having high mobility and spacing abilities do not need any web structure but should incur the costs of dispersal as well as the costs of locating mates and food. Therefore, an important point to consider is that life history diversity is created not only by the effects of host spatiotemporal stability but also by predator–prey interactions in the same habitat.

3.7 Other Variation in Spider Mites on *Sasa*

We also noticed that all three overwintering types appear in *Sasa*-inhabiting spider mites (see Table 3.3). As with life history variation, we must search for an alternate explanation for such diversity in overwintering because there is no way to adapt the host plant phenology hypothesis (Sect. 2.4) to species living on the same host plant. Furthermore, *Y. sapporensis* and *Pn. akitanus* have two overwintering stages – egg and adult female – on *Sasa*. (Gotoh 1986; Table 3.3).

How can we explain such overwintering diversity between spider mites living on *Sasa*? Addressing the conclusion first, I have no clear explanation of the reason(s) for this diversity. However, I can suggest several possibilities that may be related to this phenomenon. As pointed out by Harvey and Pagel (1991), there are great difficulties associated with studying evolutionary processes solely by between-species comparisons without accurate phylogenetic information. Because the mites inhabiting *Sasa* are categorized into various genera of Tetranychinae, such diversity in overwintering types appears to be simply a reflection of phylogenetic constraints. If so, namely that for some reason an ancient species of a certain genus had a particular overwintering stage, then such a trait is not always an ecological trait on *Sasa*.

Gutierrez and Helle (1985) have proposed several hypotheses on spider mite phylogeny on the basis of morphology and karyotype, and both Navajas et al. (1996) and Sakagami (2002) have considered the issue from a molecular perspective. Although neither is sufficiently comprehensive and both contain several inconsistencies, we may say that the differences in overwintering stages evident in *Sasa*-inhabiting mites are consistent with the major trends in their respective genera, such that it may be regarded mostly as a phylogenetic constraint: Although *Stigmaeopsis* is still not positioned in the Tetranychinae phylogeny, in accordance with Sakagami (2002), it may be placed near the genus *Tetranychus* in which all species overwinter as adult females (see Table 3.3; Veerman 1985; Niino and Sakagami, unpublished data), such that we can also consider overwintering at the adult female stage to reflect phylogeny. However, the overwintering in two stages observed for both *Y. sapporensis* and *Pn. akitanus* is an extraordinary and rare phenomenon in Tetranychinae (although *Sz. brevisetosus* also has two overwintering stages; Saito, unpublished data). *Yezonychus* is thought to be very close to *Schizotetranychus* in which all species overwinter in the egg stage; thus, we can suppose that the ancestor of *Y. sapporensis* primarily overwintered in egg stage. Therefore, I consider it plausible that overwintering adults recently evolved through ecological factors, such as predation pressure on *Sasa*. Anyhow, all we can say at the present moment is that there is no single concise explanation for the overwintering stage diversity seen in *Sasa*-inhabiting spider mites, other than an ecological null hypothesis, i.e., phylogenetic constraints. From this reasoning, the observed species can all live on *Sasa* leaves regardless of differences in overwintering type, because there may be no strong selection pressure on such a trait in this habitat.

Chapter 4
Mite Sociality

After a thorough description of mite diversity, I would like to move on to address the social evolution of mites. But to do so I first have to describe sex-related phenomena in this taxon, because sexual behavior can provide essential basic information to understand social interactions between individuals. Furthermore, a biased sex ratio is a very important prerequisite of kin selection, because it sometimes results in unbalanced relatedness between sibs (Hamilton 1972).

4.1 Diversity in Genetic System

Within the world of mites, we can find almost any kind of genetic system (genome-transporting systems; Table 4.1), i.e., diplo-diploidy, haplo-diploidy, and thelytoky (Walter 1996). Furthermore, an extraordinary system, "apparent" haploid-thelytoky, occurs in Tenuipalpidae (Table 4.2; Weeks et al. 2001). Understanding the diversity in genetic systems in mites (in all organisms for that matter) is an extremely important issue that is as yet unresolved (Bull 1983). Furthermore, there is much variation in how such genome inheritance systems are attained within each genetic system. For example, haplo-diploidy involves at least two different mechanisms, i.e., arrhenotoky and paternal genome loss (PGL), and thelytoky involves both automixy and apomixy. Furthermore, as stated by Wrensch et al. (1994), inverted meiosis, which is hypothesized in mites, is not negligible. Such differences in the mechanisms tend not to influence the number of genomes per zygote but rather greatly influence genetic variability. From the perspective of evolutionary ecology, the latter is rather important, because selection basically operates on such genetic variability.

Meanwhile, it is erroneous to say that genetic systems are equal to sex determination mechanisms. As Helle and Overmeer (1973) properly stated for haplo-diploid organisms, the statement that haploid individuals are males and females are diploid is true (but see Weeks et al. 2001), but stating that males are always haploid and females always diploid is not always true. This is a problem of causal sequence (cause-and-effect problem). In social bees having haplo-diploidy, it is well known that diploid males are produced from inbred females (Crozier 1979; Bull 1983),

Y. Saito, *Plant Mites and Sociality: Diversity and Evolution*
DOI 10.1007/978-4-431-99456-5_4, © Springer 2010

Table 4.1 Genetic systems known in Acari

Genetic system		Genome set Male	Female	Taxa in which examples are known
Diplo-diploidy		2n	2n	Ixodida
Haplo-diploidy				
	Arrhenotoky	n	2n	Prostigmata (tetranychids)
	Paternal genome loss[a]	n	2n	Mesostigmata (phytoseiids)
	Gynogenesis[b]	n	2n	Mesostigmata (dermanyssids)
Thelytoky	Haploid thelytoky		(n)	Prostigmata (tenuipalpids[c])
	Diploid thelytoky		2n	Prostigmata (tetranychids, cheyletids), Oribatida, Mesostigmata, and Ixodida

[a]Mating is necessary for female reproduction, but the male genome is eliminated after fusing with the egg nucleus

[b]Mating is necessary for the female to reproduce because egg embryo development is only activated by a sperm. However, the sperm degenerates without fusing with the egg nucleus (some times considered as thelytoky)

[c]*Brevipalpus phoenicis*, which has this type of genetic system, is controlled by microorganisms (see text; Weeks et al. 2001)

Table 4.2 Genetic systems in Tetranychidae and its related family, Tenuipalpidae

Family or subfamily	Thelytoky	Arrhenotoky[a]	Others	No. of species checked
Teytranychidae				
Bryobiinae	35%	65%	–	17
Tetranychinae	2%	98%	Tychothelytoky[b]	116
Tenuipalpidae	+	+	Haploid-thelytoky[c]	–

[a]Including species in which unfertilized oviposition and male existence were only observed

[b]In *Oligonychus ilicis*, unfertilized females reproduce males, but sometimes they also produce females (Helle and Pijnacker 1985)

[c]See Table 4.1

a phenomenon explained by the following two models. One is that two alleles at a locus determine sex, i.e., hemizygotes (haploid) and homozygotes (diploid) become males and heterozygotes becomes females (the CSD model, Bull 1983; Van Wilgenburg et al. 2006). The other is the multi-locus homo-hetero relationship, i.e., hemizygotes and homozygotes at multiple loci become males and heterozygotes at least at one locus become females (Bull 1983). Therefore, it is usually appropriate to say that hemizygotes (gametes) develop into males, but not that zygotes (diploid) always develop into females.

Another recently discovered example is *Brevipalpus phoenicis* (Tenuipalpidae, Fig. 1.1), which reproduces parthenogenetically. Weeks et al. (2001) observed the details of this species by molecular and bacteriological methods, and revealed that the "female" is actually a "male!" This paradoxical result has surprised many researchers, including myself. Weeks et al. treated *B. phoenicis* females with an antibiotic and got adult male offspring, which had been hitherto unknown in this species. It is known that microorganisms, such as *Wolbachia* spp., can manipulate host sex ratios and feminization in arthropods, but most effects are restricted to controlling sex ratio by killing one sex gamete (O'Neill et al. 1997; Stouthamer et al. 1999). The data obtained by Weeks et al. (2001) illustrated that feminization can involve bacteria other than *Wolbachia* and that there are microorganisms which can change a host's sex from "male" to "female" directly! This clearly showed that not only genetic factors (i.e., sex determination genes and homo-hetero relationship) but also parasitic microorganisms can determine sex.

For haplo-diploid spider mites, we have no information about the "strict mechanisms" of sex determination, although a few hypotheses have been provided (Helle and Overmeer 1973). In relationship to the CSD model, Saito et al. (2000b) conducted extremely intense inbreeding experiments (ten generations of mother–son mating) on *St. miscanthi* (Chap. 5) but could not observe any sex aberrations (of course, this does not deny that all such individuals were selected out before maturity), suggesting diploid males (such as drone male bees) may be nonexistent (Saito et al. 2000b). Therefore, it is difficult to hypothesize sex determination mechanisms as homo-hetero zygotic relationships (including the CSD model). By observing sex aberrations in an inbred line of *Tt. urticae*, Van Eydhoven and Helle (1966) presented another hint. These inbred individuals were sex mosaics, which were described as "giant males" and "intersexes." However, Helle and Pijnacker (1985) stated that genetic studies with the strain producing the sex aberrants did not provide information on the genetics of sex determination. From such results, they supposed that the female sex must be controlled by the karyotype-doubling, because polyploid individuals were supposed to be sex aberrants, suggesting that a degree of ploidy higher than diploidy seems to increase the effect of male-determining factors. Thus, one of the more plausible mechanisms concerns DNA-level modification (e.g., methylation) as known for a green alga (Nishiyama et al. 2004), where the maternal gamete is modified but the paternal gamete is not. If such parental sex-specific modification of DNA at the "gamete level" occurs and the modified gamete controls male sex, then the heterozygote of nonmodified and modified gametes becomes a female and the modified gamete (hemizygote) becomes a male. Whether such mechanisms operate in the sex determination of spider mites as well as other haplo-diploid mite species (e.g., Phytoseiidae) must be studied at the gene level in the future.

Indeed, the variance in genetic systems in the Acari may provide other perspectives in ecology, behavior, and evolution. I address these topics repeatedly in the later sections on inbreeding depression and social evolution.

4.2 Sexual Behavior

Males show at least three kinds of "mate guarding." One is a monogamous system, where males guard or defend partner females from pair formation to development of offspring, as is known in many birds and mammals. The second is called postcopulatory mate guarding and is known in damselflies and dragonflies (e.g., Tsubaki et al. 1994), where males guard partner females from just after copulation to egg laying. Postcopulatory mate guarding would evolve when the last male precedence (eggs are fertilized by the sperm of the last male to mate) is sufficiently high; e.g., male damselflies can remove the sperm of previous males and replace it with their own by use of their genitalia (Waage 1979).

The other form is called precopulatory mate guarding (abbreviated as pre-copula; Ridley 1983) and is known from several taxa, such as toads, amphipods, and mites. In contrast to postcopulatory mate guarding, pre-copula would evolve when the first male can exclusively fertilize a female (Ridley 1985; Yasui 1988). For example, amphipod females can mate just after molting (note that they repeatedly molt after maturity and become receptive just after each molt), such that males have to find such "soon-to-be-available" females to mate. Therefore, they have to guard unmated females so as to mate them just after molting and maturity. Although no such mating restriction from female morphology has been known in mites (i.e., females can mate for a long time after maturity), several species show precopulatory guarding (Potter et al. 1976; Yasui 1988). Yasui (1988) reported that the males of *Macrocheles muscaedomesticae* (Acari, Dermanyssina) having haplo-diploidy also show precopulatory guarding. By using mating experiments with genetic markers, he was able to show that first male precedence is very high (99.8%). He discussed not only external morphological constraints as in amphipods, but also several other factors that make sperm exchange by later males impossible, thereby guaranteeing a high probability that eggs are fertilized by the sperm of the first male (called P1). Precopulatory guarding must have evolved to assure "sperm" priority. In addition to this, he observed the mating behavior and P1 in *Parasitus fimetorum* (Acari, Parasitina, diplo-diploidy), which revealed that in this species males never show precopulatory guarding and that P1 is low when females mate repeatedly with a time lag. This result is consistent with his hypothesis for *M. muscaedomesticae*.

Precopulatory guarding is also known in many spider mite species (Fig. 4.1) including *Ap. corpuzae*, *St. longus*, and *Tt. urticae* (Potter et al. 1976). Females of *St. longus* can mate for a long time after maturity if they are still in an uninseminated state (Saito 1987); then there is no morphological restriction in female copulation. On the other hand, it is known that the first males exclusively fertilize all the eggs of partner females in *Tt. urticae* (Boudreaux and Dosse 1963; Helle 1967), such that Yasui's (1988) proposition may be also applicable to spider mites.

Meanwhile, the length of the teleiochrysalis stage (third quiescent stage just before molting to the adult) of spider mite is very long in comparison with *M. muscaedomesticae*. Thus a male strategy for determining how long they should guard is to determine at what "age" the third quiescent stage females are (Grafen and Ridley 1983). From the perspective of males, the best strategy is to shorten the guarding time before mating because males who can minimize guarding time must increase their chances of finding other females. Potter et al. (1976) observed whether males tend to guard older third quiescent stage females that will soon molt. They showed that males tend to guard third quiescent stage females distributed over the experimental leaf surface more in the last period of the experiment. They concluded that males have a

Aponychus corpuzae

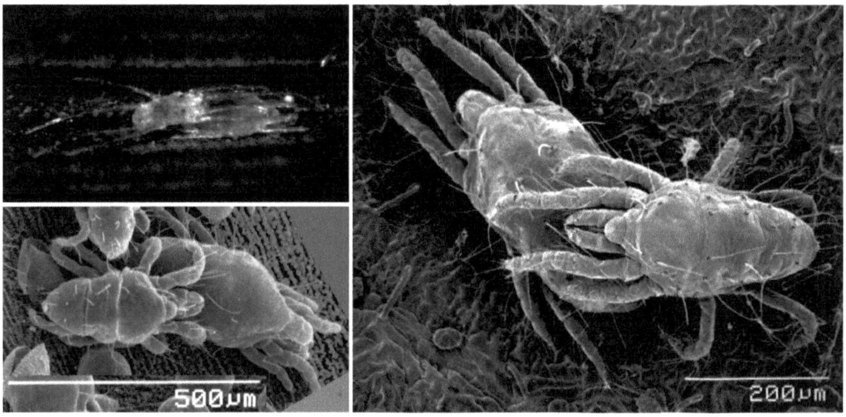

Stigmaeopsis longus *Tetranychus urticae*

Fig. 4.1 Precopulatory mate guarding in three spider mite species

strong tendency to guard older quiescent females (just before ecdysis). However, they introduced active deutonymphs onto the experimental arena to establish different-aged quiescent females. Thus, their findings do not always mean the existence of male strategy to guard older quiescent females to save guarding time; this is because the guarding frequency must be also a function of the length of time quiescent females are exposed to males. If we wish to solve this question in a clearer manner, different-aged third quiescent stage females must be simultaneously provided to a single male (the reason for using a single male is to minimize the effect of the male-to-male competition observed by Potter et al. 1976); then the age of any female guarded by the single male could be observed. Therefore, we do not know whether spider mite males actually possess any such strategy as theoretically predicted by Grafen and Ridley (1983). I believe that such a study would be very relevant.

We do have a detailed understanding of the variation in pre-copula behavior and male-to-male competition behavior for securing mates between spider mite species; e.g., *Tt. urticae* males use stylets and saliva (Fig. 4.2), *Pn. citri* males fight each other intensely by using their forelegs in a boxer-like manner (also sometimes using stylets and saliva; Fig. 4.3), and *St. miscanthi* males use their stylets to cannibalize opponents (Chap. 6; Saito 1990c). Because of the difficulty in evaluating the intensity of male-to-male combat, such differences in male behavior have mostly been ignored in comparative behavioral studies so far (but see Sect. 6.1).

On the other hand, Radwan (1993, 1995) reported a very interesting phenomenon in acarid mites. Male dimorphism is known in many acarid species. The "fighter" males of *Sancassania* (*Caloglyphus*) *berlesei* are typically characterized by a thickened and sharply terminating third pair of legs, whereas "scrambler" males have no such morphology and are benign. The "fighter" males frequently kill rivals using the thickened pair of legs. Because males can be triggered to become fighters by decreasing the density of these mites, the variation in fighter expression in this mite

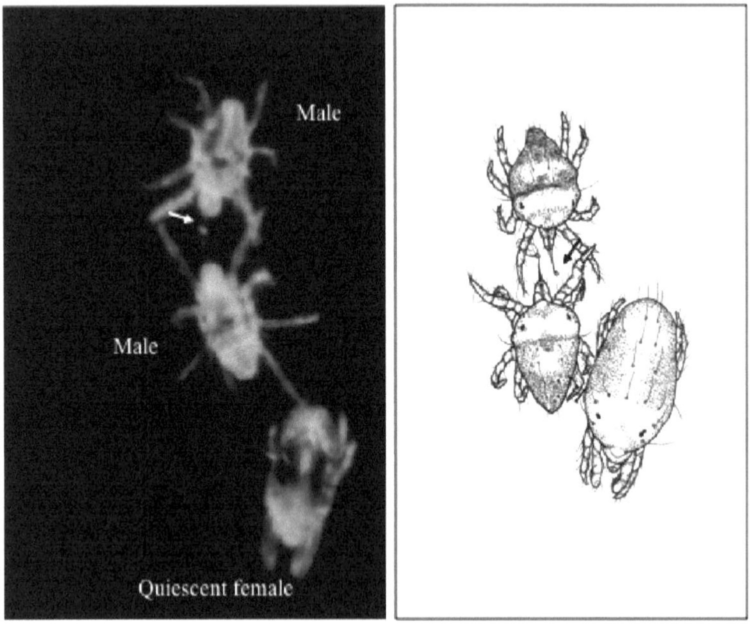

Fig. 4.2 Male-to-male combat in *Tt. urticae*. On the *left* is a photograph from a 16 mm movie and on the *right* is a drawing based on the movie. *Arrows* show saliva secreted on stylets

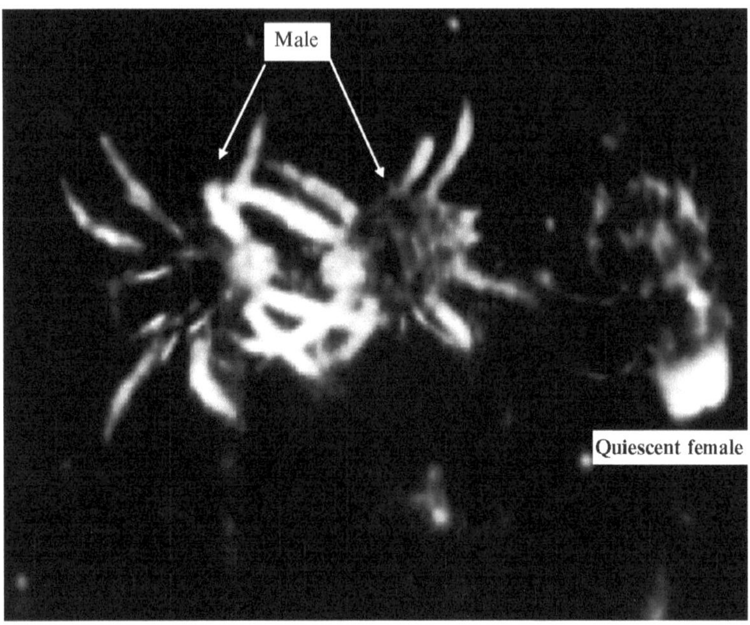

Fig. 4.3 Combat-like boxing in *Panonychus (Pn.) citri* males. Leg movement was so swift that their first legs were filmed as a two-fold thickness (filmed in 16 mm format by Saito)

is caused by the density conditions under which they developed. He supposed that the fitness benefits of the alternative tactics were that at low colony densities fighter males are likely to have an advantage over scramblers (benign), because they are able to kill all their rival males and dominate a group of females (Radwan 1993). At high colony densities, on the other hand, fighter males would derive no advantages because of the high cost (frequency) of fighting (if all fighters fought together then males would have little time to mate). Using this species, Tomkins et al. (2004) tested the existence of status-dependent evolutionary stable strategy (ESS) and showed that this mite species is quite appropriate for testing such a general theory in evolutionary ecology. These are undoubtedly nice examples for demonstrating the suitability of mites as materials for testing general theories in evolutionary ecology and sociobiology. However, I have a further question about why such a strategy occurs exclusively in acarid mite groups because I also observed a kind of mixed strategy in male-to-male antagonism in a spider mite that does not show any male dimorphisms (Sect. 6.5) within a population. I think that identifying the factors which cause such high male-to-male antagonism in animals and why some animals evolved dimorphic males and others did not are important problems that will soon be challenged by utilizing these mite groups.

Sexual behavior between males and females has received much attention in mites, because there is so much variation in the mode of sperm transfer. According to Walter and Proctor (1999), almost all known sperm transfer modes are listed in the Acari. Therefore, it is difficult to describe them one by one. Here I only want to introduce several examples and recommend any reader wishing more detailed information to check Walter and Proctor's (1999) fascinating review. All spider mite males have an aedeagus (penis) and use it to directly transfer sperm. If the sperm's path inside the female spider mite involves traveling through the hemocoele, then first male sperm precedence (P1) is thought to be caused by the traveling mode (this traveling process was also hypothesized in *M. muscaedomesticae*) (Walter and Proctor 1999). Other fascinating studies on intersexual behavior appear in Tilszer et al. (2006) and Prokop et al. (2007). Tilszer et al. (2006) used an acarid mite species, *Rizhoglyphus robini*, to show how sexual conflict decreases female fitness. They revealed that the decrease in harmfulness (to partner females) of monogamous males was accompanied by a decline in reproductive competitiveness by thirty seven generations of selection lines either enforced monogamy or polygamy. Prokop et al. (2007) discovered that the female offspring obtained by mating with older males show significantly decreased fitness, probably through the accumulation of deleterious mutations in the male germline. These are also excellent examples showing the appropriateness of mites as materials for testing general ecological theories, as reviewed by Radwan (2007).

One of the interesting phenomena related to sexual behavior in spider mites underlies the large variation in the size (and/or bending direction) of the aedeagus between species. From the point of view of the taxonomist, the spider mite aedeagus is relatively simple in shape and not so distinctive between related species, such that it is only used as a secondary character for species description. However, if we switch our perspective to the aedeagus size differences between species, then its length varies from about 13 μm in *Tt. urticae* to about 67 μm in *Eotetranychus tiliarium*

(Ehara and Shinkaji 1975). If we compare the length of the aedeagus to male body size, in the former it is only about 2% but in the latter it reaches about 23% of body length. There is no behavioral information about why such variation exists, but I believe that it is some reflection of a mating strategy.

4.3 Sex Ratio

It is well known that the sex ratio of mites varies considerably. Hamilton (1967) first addressed the evolutionary reason why biased sex ratios evolved in various organisms, even though a 1:1 ratio is primarily an evolutionary stable strategy as shown by Fisher (1930). Fisher (1930) solved the reason why sex ratios attain evenness in both sexes from the theory of frequency-dependent selection: Under random mating conditions, as the frequency of one sex increases, mothers who produce more offspring of the rarer sex will get more advantage (many grandchildren). Thus the sex ratio = 1/2 attains an evolutionary stable strategy (ESS). Hamilton (1967) considered this principle as "a starting point of the sex ratio problem." Then he considered "nonrandom mating" by introducing spatially structured mating populations (Box 4.1). By the use of a game theoretical approach, he showed that the sex ratio will become biased toward the female sex if the following conditions

Box 4.1 Basic mating structure of Hamilton's local mate competition (LMC) theory

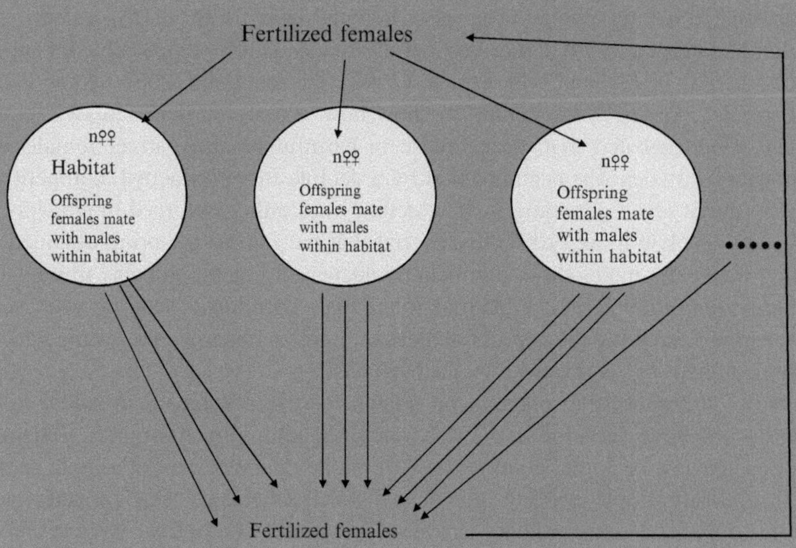

are met. A fertilized female establishes a patch with $n-1$ females (unrelated) and reproduces offspring within the patch (there are no immigrants). Offspring (males and females) mate within the maternal patch, then disperse to establish a new patch. There is no difference in the mortality of offspring females. Under these conditions, the optimal sex ratio (ESS sex ratio, SR*) of offspring produced by a female will change depending upon n, and it can be calculated as $SR^* = (n-1)/2n$ both for diplo-diploids and to a good approximation also for haplo-diploids, if there is no relatedness between foundresses (Hamilton 1967; Herre 1985; Werren and Colwell 1987). This is the famous local mate competition (LMC) theory for the evolution of female-biased sex ratios. Hamilton (1967) listed several animal taxa as candidates for fitting his LMC model. Mites are one such taxon in which biased sex ratios have frequently been reported (Table 4.3).

After Hamilton (1967), the biased sex ratios observed in animals have been considered as evidence for the LMC theory. However, the biased sex ratio by LMC only has explanatory power if we have sufficient information about the mating structure of the target species. For many organisms except some parasitic wasps (West et al. 2001), it is hard to determine in the field how many females establish a patch, whether there is dispersal between patches, how many generations the patch can serve, and so on. Therefore, examples that clearly show a biased sex ratio having evolved by LMC are scarce, except for parasitic wasps.

Table 4.3 Sex ratio observed in spider mites (at maturity)

Species	Male ratio	Sources
Aponychus corpuzae	0.33	Saito and Ueno 1979
Eotetranychus uncatus	0.16	Gotoh 1987a
Eotetranychus tiliarium	0.21	Gotoh 1987a
Oligonychus castaneae (ununguis)	0.47	Saito 1979a,b
Panonychus (Sasanychus) akitanus	0.11	Gotoh 1987b
Panonychus citri	0.25	Saito 1979a,b
Schizotetranychus bambusae	0.21	Saito, unpublished data
Schizotetranychus cercidiphylli	0.25	Gotoh 1989
Schizotetranychus lespedezae	0.28	Gotoh 1989
Schizotetranychus schizopus	0.17	Gotoh 1989
Stigmaeopsis celarius	0.28	Mori 2000
Stigmaeopsis longus	0.14	Saito and Takahashi 1982
Stigmaeopsis longus	0.15	Saito and Takahashi 1982
Stigmaeopsis miscanthi	0.15[a]	Sato et al. 2000
Stigmaeopsis saharai	0.21	Mori 2000
Stigmaeopsis takahashii	0.21	Mori 2000
Stigmaeopsis takahashii	0.16	Saito and Takahashi 1982
Tetranychus cinnabarinus	0.43	Hazan et al. 1973
Tetranychus desertorum	0.33	Nickel 1960
Tetranychus urticae	0.40	Mitchell 1973
Tetranychus urticae	0.32	Saito 1979a

Data are sex ratio at maturity under 20–25°C
[a]Median of eight populations

Box 4.2

As pointed out by Hardy and Mayhew (1998), there is a well-known negative correlation between brood size and (male) sex ratio in Bethylidae. If a patch is occupied by a single fertilized female only, there will be an extreme number of males; i.e., a single male must be expected by the LMC model (note that the equation predicts no males, but this is impossible for bisexual organisms). In such an extreme case, the female fecundity must greatly influence the sex ratio observed, because a single male is always produced regardless of fecundity. Fecundity increase must increase female offspring and then decrease male ratio. Therefore, we cannot neglect the effect of fecundity when we attempt to test the LMC model.

As shown in Table 4.3, there is much variation in the sex ratios of spider mites and almost all are more or less female biased. It was very difficult to determine the "primary sex" ratio of spider mites, although now it is possible by the use of molecular-level methods (but few attempts have been made). Therefore, the data shown in Table 4.3 are all progeny sex ratio at maturity (note that the offspring survival rate is sufficiently high to be considered them an approximation of primary sex ratio as well). To prove LMC in spider mites, two studies have been conducted that revealed there are phenomena which are consistent with LMC predictions. Roeder (1992) showed that *Tt. urticae* females produced more males when they were reared under high-density conditions, although the numbers of eggs laid per female changed with female density (Box 4.2). Sato and Saito (2007a) also reported that *St. longus* and *St. miscanthi* females (both living in nests with cooperative sociality) changed their sex ratios according to the nest foundress density, i.e., singly foundresses produce significantly fewer males than two or more foundresses. They simultaneously showed that there is no difference in female fecundity between different foundress densities, such that they could conclude that the results showed the existence of LMC. However, because the observed sex ratios were higher than those expected from the LMC model for haplo-diploidy (Taylor and Bulmer 1980), the question of why there is "surplus female-biased sex ratio" in these species arose.

The foregoing two studies were conducted under experimental conditions, such that they could only show the fact that foundress density changes offspring sex ratio. As stated before, the most important prerequisite of Hamilton's LMC is the presence of spatially structured habitats in which mating takes place (Hamilton 1979; Taylor and Bulmer 1980). In nature, as we have no accurate information about mating population structures in space and time for mite species, we must devote much more attention if we intend to apply the LMC model to these spider mite species (Box 4.3). This is a problem common to the other reasonings of sex ratio variation in mites mentioned later.

There are two other models that try to explain biased sex ratio in haplo-diploid mites. Nagelkerke and Sabelis (1996) adopted the "haystack model" of Maynard

Box 4.3

In relationship to these, Roeder et al. (1996) experimentally showed that sex ratio shifts in response to the relatedness between foundresses when the number of foundresses is equal. They suspected that such immediate responses in sex ratio suggested that *Tetranychus urticae* female has some method of discrimination between kin and non-kin.

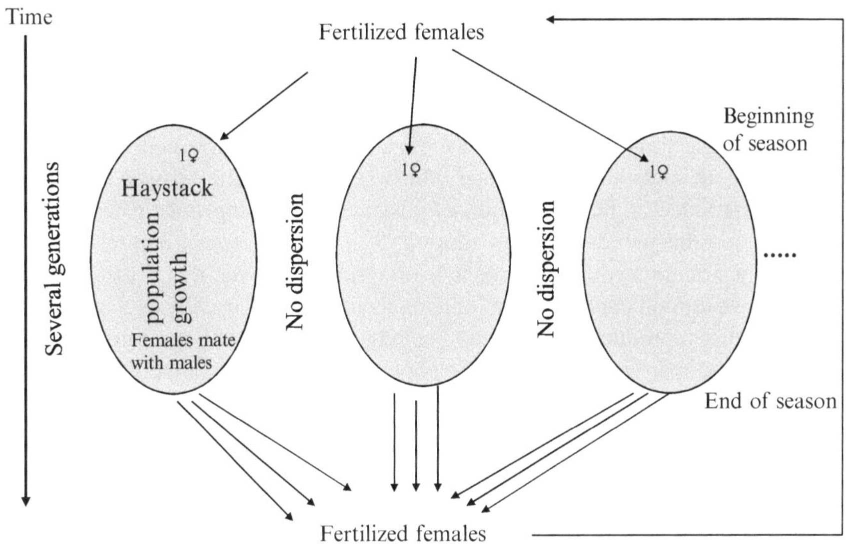

Fig. 4.4 Basic structure of haystack model

Smith (1964) for explaining the biased sex ratio of phytoseiid mites (haplo-diploidy). The haystack model is also explained in mite social evolution in Sect. 4.7, such that I only describe it here in essence. The basic conditions of the model are as follows (Fig. 4.4): Haystacks refer to resources that accumulate or appear once in the beginning of a season (start of animal reproduction). These resources are distributed patchily and are altogether isolated. Animals that were inseminated within their natal patches in the previous season migrate randomly and singly into these new patches and reproduce there for several generations. There is no dispersion of individuals between patches (thus they are isolated together) after immigration, and all the haystacks deteriorate (become exhausted) simultaneously at the end of the season. Just before the disappearance of the haystacks, all adult females mate within their natal haystacks and disperse (males never disperse). Then the life cycle starts again at the beginning. There is no difference in immigration success between

Box 4.4

The haystack model itself is an individual-based (1-locus and 2-allele) model. Maynard Smith (1964) presented these conditions as improbable assumptions by use of this model. In the 1960s, there was much debate about the existence of group selection, i.e., organisms have evolved to maintain their group, e.g., flock, population, or species. The most famous and influential opinion was provided by Wynne-Edwards (1962). However, as stressed by Williams (1966), selection basically operates on individuals, not on groups, and situations where group selection operates are uncommon. The haystack model has hypothesized such conditions, but as Maynard Smith stressed himself, the necessary conditions are unrealistic.

individuals regardless of the haystacks from which they originated. Under these assumptions, animals should produce which sex ratios? The answer is obvious. Females reproduce as many females as possible by keeping male offspring to a minimum. In other words, females adopt an r_m (intrinsic rate of natural increase) maximizing strategy by increasing female offspring. However, this situation is very curious, if we remind ourselves that individuals in a species usually exchange genes through sexual reproduction (thus they belong to a species in the sense of Mayr 1963). Because there is no interbreeding between haystack populations, each haystack population could be considered a different species or thelytokous organism in which sex ratio strategy never evolves. Therefore, as stated by Maynard Smith (1964), these conditions must be not realized in a species with sex (Box 4.4), such that it is difficult to utilize this model in its original state to explain female-biased sex ratio in mites.

Then, Nagelkerke (1993) relaxed several conditions of the original haystack model as follows (some of them were also analyzed by Maynard Smith 1964) and simulated the optimal sex ratio. The number of immigrant females per haystack was changed from 1 to n (more than 1) individuals (non-kin), and the maximum number of individuals attainable at the last generation (=dispersal generation) was restricted. If sex ratio is fixed for an individual's traits (inflexible), the increase of n gradually decreases the female ratio in the population, but if sex ratio is flexible for females at each generation, it immediately reaches evenness (1/2) other than $n=1$ (Nagelkerke 1993). Furthermore, he attempted to change the mating structure within a haystack and found out that subdividing haystacks in one-generation mating groups promotes the female bias and the unbeatable (optimal) sex ratio can change radically with generations within haystacks.

I think that a distinct difference between the LMC model and the haystack with unlimited growth model (Nagelkerke 1993) exists only in the number of generations in a patch; namely, the former model assumed that only one generation (from foundress reproduction to offspring development and mating) passes in a patch and the latter that several generations pass there. Therefore, the latter and more complex models should be considered extensions of LMC. However, it is also difficult to know the real state of habitat structure as well as mating structure in the field, such that all are equally plausible for explaining biased-sex ratios in mites.

The foregoing reasoning about sex ratio evolution is commonly hypothesized from individual (gene-to-gene) level selection. On the other hand, there is a model in which group selection is hypothesized. Wrensch (1993) presented the "r_m–S_m" model for explaining female-biased sex ratio in haplo-diploid mites as well as "colonizing species." A colonizing species is synonymous to *r*-selected species (or *r*-strategists), which use unpredictable habitats (resources) in an opportunistic manner (Sect. 2.2). In such biofaces, she said that group selection, i.e., colony-level survival, must be important, because high r_m and high genetic variability (denoted as S_m; especially good gene sets in a female for survival during dispersal and colonizing in new habitats) must be favored to produce success. To realize this, colonies having high r_m in the early episodes of colonization will be at an advantage because of how fast and how numerous the colonized populations grow in the time-limited resources. This status may be realized by highly female-biased sex ratios (note that no individual-level selection is assumed there). On the other hand, in the late episodes of colonization when resources are nearly exploited, "colonies" (thus "groups") producing many males must have an advantage for the following reasons: First, haploid males are hemizygotes, such that all male genes are directly exposed to the environment without any masking effect from the dominance-recessive relationship. If a "population" produces many males, bad genes (mutated during population growth) in the colony must be immediately and effectively selected out through males, and only sound males may live in the colony (Wrensch 1993). This is just as in the story hypothesized by Atmer (1991), who emphasized the role of haploid males as a "purge" mechanism for discarding bad (deleterious mutation) genes (although I think this idea is flawed). Furthermore, she hypothesized that the variability of gene combinations at the population level (i.e., all dispersing females + sperm) must increase the opportunity to reach other habitats and to establish new colonies there. Therefore a "male-biased sex ratio" in the later phases of colonizing should be advantageous. From the above, Wrensch (1993) provided a new idea, an "r_m–S_m" strategy at population level in spider mites. Her model basically arose from the sex ratio determination mechanisms observed in a spider mite species, *Tt. urticae* (Young et al. 1986). In this species, the sex ratio is determined by using only a sufficient number of sperm to fertilize a fixed fraction of eggs per day, suggesting the females "do not control sex ratio," but the number of daughters. In other words:

> … fertilization pattern is independent of progeny number. Because this control pattern serves to produce more daughters, whatever the total fecundity or degree of heterogeneity among individual progeny sex ratios, it maximizes the NUMBER of daughters in a progeny, not the FRACTION of daughters. Thus, control of daily sex ratio does not conform to either the assumptions of predictions of optimality theory of sex ratio…,

Wrensch (1993) said.

It is worthwhile to move our focus to daily sex ratios in iteroparous spider mites and to the importance of the population process, but I think that her "r_m–S_m" model of colonizing species should be regarded as one of the possibilities. Group selection or multilevel selection is considered to be a "hyper-hypothesis" that inevitably involves more complex assumptions (in her model, no individual selection was assumed) than individual-based selection (Williams 1966), such that irrefutable evidence other than individual selection is necessary, as provided by Tsuji (1990) in ants.

As mentioned before and again later, there are many spider mite species having *K*-strategies, such as *Stigmaeopsis* species on evergreen bamboo plants (living in very stable habitats; Saito and Ueno 1979). However, the sex ratio of these species is strongly female biased (see Table 4.3). I have already discussed in this book that these species do not adopt the strategy (*r*-strategy) of colonizing species, such that their female-biased sex ratios cannot be explained from the "r_m–S_m" model.

Another theory of female-biased sex ratio was recently presented. The parasitoid wasp *Melittobia australica* (Eulophidae) is known to show a more extreme female-biased sex ratio than predicted by the usual LMC and inbreeding models (Abe et al. 2003a,b). Abe et al. (2003a,b) proposed a new idea: that a mother should save her male progeny if there is mortal competition between males. In *St. miscanthi*, male-to-male competition is extremely high (Chap. 6), and the female-biased sex ratio in this species seems to fit well with their theory. However, there is no male-to-male competition in *St. longus* (Saito 1990c), even though the sex ratio is skewed to almost the same extent as in the highly aggressive *St. miscanthi*. Therefore, their male-to-male competition theory may not fit the present case.

As such, I have to state here that "evolutionary" factors capable of biasing the sex ratios of mites are still open problems for future study, although several important models have been proposed. For spider mites in particular, we are now wondering why the sex ratios of *Stigmaeopsis* species in which the number of foundresses and that of generations per patch (nest; Sect. 3.5) vary are constantly female biased (see Table 4.3), even though they have the ability to respond to foundress density as predicted by LMC (Sato and Saito 2007a).

In relation to the above, we have to focus our attention on control mechanisms other than the genes of the target species, such as sex control by parasitic microorganisms as known in many insects and Tenuipalpidae (Weeks et al. 2001; Sato and Saito 2007a).

4.4 Sociality in *Stigmaeopsis*

Mori and Saito (2005) suggested that the nest size is correlated with the effect of nest defense (counterattack). Twenty-two years ago, I reported counterattack behavior by females and males in *St. longus* (known as the long seta form of *Schizotetranychus celarius* at that time), i.e., large nest builders, against a specific predator species, *Tp. bambusae*, as mentioned previously (Sect. 3.5, Saito 1986a). This finding is a kind of social behavior that must be considered to arise from biparental care. However, I received a criticism from Crozier (1985) about this reasoning:

Box 4.5

There was a reason why I could get his comments in the book *Spider Mites. Their Biology and Control* (eds. W. Helle and M.W. Sabelis, 1985) before the publications of my original studies (Saito 1986a,b); when I visited Prof. Helle to consult about my part in the manuscript after the 17th International Congress of Entomology in Hamburg (1984), I told him of my recent discovery about mite parental care, and he recommended that I send my manuscript immediately to Prof. Crozier, who was also a contributor to the book. Thus I had an opportunity to get his comments in this influential book.

> The sociobiology of spider mites has received a significant boost from the recent discovery (Saito 1986a) that adults of *Schizotetranychus celarius* (which lives in a multi-chambered web nest) eject or kill invading predatory phytoseiid mites, thus protecting immature mites in the nest (which are not always their own offspring). If further experiments show that these adults are acting to protect the immatures rather than just themselves, then the casc would clearly be one of parental care

(Box 4.5). Actually, showing whether *St. longus* (*Sz. celarius*) females and males attack predators to defend their offspring or simply as self-defense posed a difficult problem. Although I briefly wrote the reasons why I believe the former in a later publication (Saito 1986a), a more detailed explanation is as follows.

In behavioral studies, researchers often encounter such kinds of questions when they focus on "altruistic" traits. For example, when a raptorial bird approaches a prey nest containing several chicks, a prey parent bird often flies away from the nest and begins a distraction display. This behavior has the effect of startling the predator and diverting its attention away from the nest. Why do we immediately regard such behavior as parental care (chick defense) behavior without any hesitation? The reasoning must be as follows: Parent birds can escape from the predator if they need to (in other words, the predator is not dangerous to the parents, but dangerous only to their chick(s)). Even in such a situation, parents perform a behavioral trait that must increase the fitness of their offspring. As such, even though the researchers cannot understand the direct "motivation" of the parent birds, they can understand that the function of such parental behavior is "parental defense." Of course, if we intend to say such a behavioral trait evolved through natural selection (raptor predator), we have to evaluate the effect of this behavior on the fitness of offspring.

Thus, if I intend to define the behavior of *St. longus* parents as biparental care, I have to at least have shown evidence such as in the case of the parent birds just mentioned, although I had evidence of offspring fitness gain by this behavior. One of the important criteria for stating this may come from predator–prey interactions as observed in Saito (1986a,b). I used *Tp. bambusae* larvae as "predators" for the behavioral observations. Initially I had no idea about the concept of parental care and simply wanted to examine the prey adult behavior against the

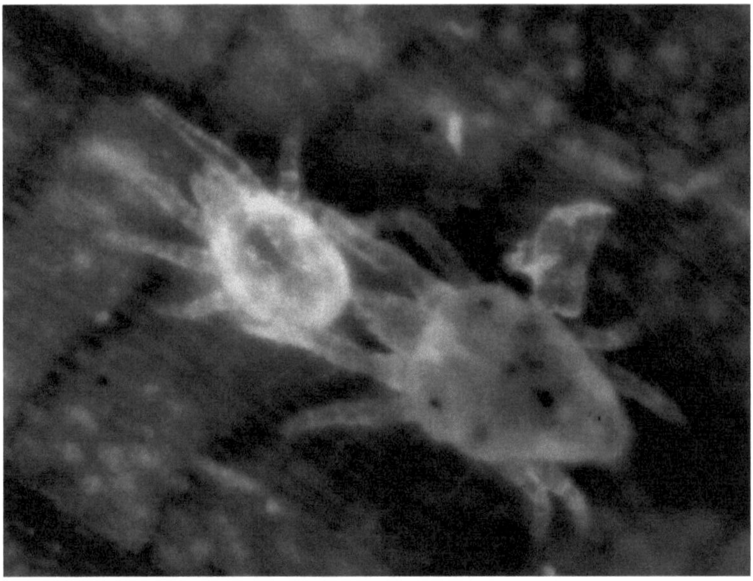

Fig. 4.5 *St. longus* male counterattacking against predator larva

predator. By using the weakest predator stage, I simply intended to get clearer evidence of counterattack. After receiving Crozier's (1985) criticism, I felt very fortunate he had mentioned the concept of parental care. As stated before (Sect. 2.5), *Tp. bambusae* larvae are nonfeeding (Chittenden and Saito 2001), such that this predator stage never attacks the prey. Therefore, there is no need for the prey adults to attack the predator, at least until it develops into a feeding stage. However, the prey adults (parents) very strongly attack the predator larvae and frequently kill them (Saito 1986a; Fig. 4.5). Furthermore, I observed that whenever an adult female predator invaded a nest where two prey females or two males resided, one of the parent females soon abandoned its offspring and nest while the other was preyed on by the predator, but both the parent males attacked and were most often killed by the predator (Fig. 4.6). There was a chance for one adult male to escape from their nest when the other male fell victim to the predator. So why doesn't he take it? These observations clearly showed that the males attack the predators not to serve themselves, but to serve their nest and offspring, i.e., they really show biparental defense! Soon after, Saito (1986a,b, 1990c), Zhang (2002), and Mori and Saito (2004) discovered similar counterattack behavior in *St. miscanthi*, *St. nanjingensis*, and *St. celarius*, respectively, as mentioned earlier (Table 4.4; see Box 3.7; Sect. 3.5).

Another origin of social behavior in *Stigmaeopsis* comes from cooperative nest building and usage. In Fig. 4.7, an example of the nesting pattern of *St. longus* on *Sasa* in the field is shown. Several important features are observed. First, this nest consisted of 15 nest cells that were progressively added by nest members, one by

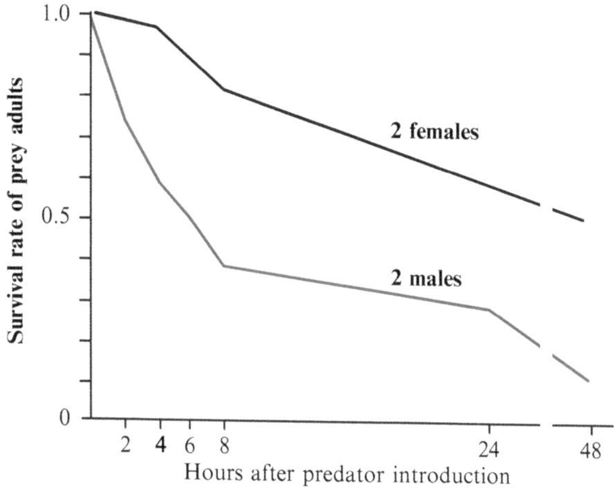

Fig. 4.6 Predation by *Typhlodromus (Tp.) bambusae* on *St. longus* females and males

Table 4.4 Sociality in *Stigmaeopsis* species

Stigmaeopsis species[a]	Nest size (=group size) for interaction	Generation overlapping	Counterattack efficiency	Waste management	Sociality level
longus	Largest	3 generations	Highest	Outside nest by chemical cues	Communal social
miscanthi HG	Large	2–3 generations	High	Inside nest by chemical cues	Communal social
miscanthi LW	Large	2–3 generations	High	Inside nest by chemical cues	Communal social
nanjingensis	Medium–large	2–3 generations	High	Inside nest; cues unknown	Communal social?
celarius	Medium	1–2 generations	Medium	Outside nest by tactile cues	Subsocial
takahashii	Small	Short	Low	Outside nest by tactile cues	Subsocial
tenuinidus	Small	Short	?	Outside nest, cues unknown	Primitive subsocial
saharai	Smallest	Short	No effect	Outside nest by tactile cues	Primitive subsocial

[a]Development of sociality increases from bottom to top of list of species
HG and LW, see Sect. 6.4

one. If a new nest cell was added after an older cell was exhausted, nest individuals ceased using the older cell. However, this figure shows that all nest cells of a continuous (united) nest are more or less utilized by nest members. Such large, united nests (about 15 cm in length) are thought to be a result of *St. longus* longevity and philopatry, and the oldest nest cell must have been built at least 1 month

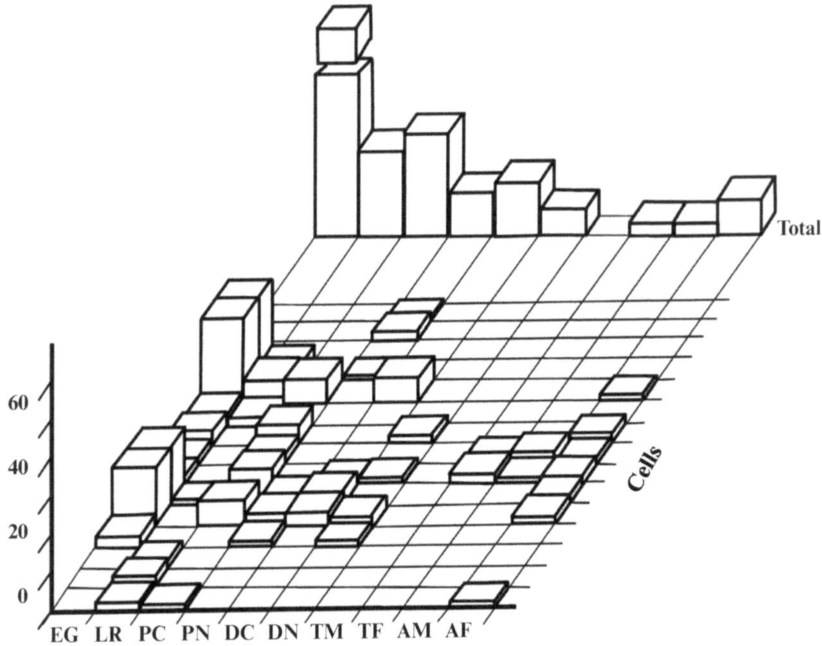

Fig. 4.7 Age structure and number of *St. longus* in all cells of united nests. *EG* egg, *LR* larva, *PC* protochrysalis, *PN* protonymph, *DC* deutochrysalis, *DN* deutonymph, *TM* teleiochrysalis male, *TF* teleiochrysalis female, *AM* adult male, *AF* adult female (from Saito 1986a)

previously (note that the generational turnover occurs about every 15 days under 25°C; Saito and Ueno 1979). This observation means that some nest members had already commenced new nest cell construction before the deterioration of the previous one. If they have no cooperative tendencies, nest members will not begin building a new nest cell before the old one has deteriorated, because to build a new nest cell they have to venture outside the nest where the danger of predation is greatly increased (cf. Sect. 3.1). Furthermore, several females sometimes initiate new *St. longus* and *St. miscanthi* nests during the season when new leaves develop (Saito 1987; Saito, unpublished data). These facts suggest that there is some merit in cooperative nest construction.

A fascinating element of the social organization of *Stigmaeopsis* species is its fecal manipulation behavior. As shown in Fig. 3.23, we observed one or two brownish grains on the leaf surface covered by the nest (most exist at the ends of the major nest axis). Behavioral observations showed these to be accumulations of feces deposited by nest members (Saito 1997; Sato et al. 2003). Because mites feed off the host plant juices, their feces are watery. Touching these watery feces is very dangerous for the mites, because they can easily become trapped by the feces and die (Sect. 2.3). If the nest members defecate everywhere in a nest, they may face great difficulties. Therefore, this fecal depositing behavior strongly suggests that the nest members perform some cooperative nest sanitation behavior. In other words, they maintain public lavatories within their nests.

Sato et al. (2003) and Sato and Saito (2006) determined the mechanisms by which *Stigmaeopsis* mites select the defecation site. Under the hypothesis that mites use tactile stimuli and/or chemical cues to deposit their feces at particular sites, we designed manipulation experiments to find out whether female mites use these two cues to maintain their defecation sites: if females use only the tactile stimulus from the nest structure to maintain the defecation sites, they should always defecate in the same places regardless of the location of the fecal pile. And if they depend upon olfactory stimuli from their feces, they should deposit their feces at the site of a relocated fecal pile, or an odor extract thereof.

First, we determined whether all nest members tended to defecate at specific sites. The results clearly showed that nest members of *St. longus* and *St. miscanthi* (*St. nanjingensis* in China also does so; Zhang and Saito, unpublished data) defecate at only "one site" inside or outside the nest. Furthermore the waste management behavior is maintained by two simple rules. First, mites defecate near the nest entrances if no volatile chemical cues are available, and second, when chemical cues are available from feces deposited previously, *St. longus* and *St. miscanthi* defecate at this site (see Fig. 3.23).

According to these rules, a foundress that does not have access to chemical cues fixes the defecation site by tactile stimuli just after constructing the framework of her nest. In this nesting and defecation process, Sato et al. (2003) discovered an interesting behavior in foundresses. Just after the introduction of a *St. miscanthi* female onto a new Chinese silvergrass leaf, there is no nest web, such that the female does not defecate for a long time. After the framework of the nest has been completed, she defecates at an entrance of her nest. Thereafter, the foundress, as well as any other individuals that may subsequently join the nest, deposit their feces at the fixed site based on odor cues (Fig. 4.8). This observation suggests that the foundress delays its defecation until the completion of the nest framework! Furthermore, Sato and Saito (2006) discovered that the chemical cues originate from chemicals produced by host plants.

In addition, if these mites can use tactile stimuli for such behavior, why have they additionally developed chemical cues? The nest has a simple tunnel-like structure, with two similarly shaped entrances. If the entrance structure is the only tactile cue for recognizing the defecation site, there will inevitably be two defecation sites. Because the nests are often extended continuously at either end (united nests) in *St. longus* and *St. miscanthi*, double defecation sites may disturb the nest extension. This explanation is strongly supported by the fact that *St. takahashii* and *St. saharai* (both small nest builders) build separate nests and defecate on both sides of the nest entrances (Sato et al. 2008) by tactile stimuli only. No chemical cues are employed. We believe the sophisticated mechanisms that enable *Stigmaeopsis* mites to deposit their feces at particular places strongly suggest that their defecation behavior evolved under strong natural selection.

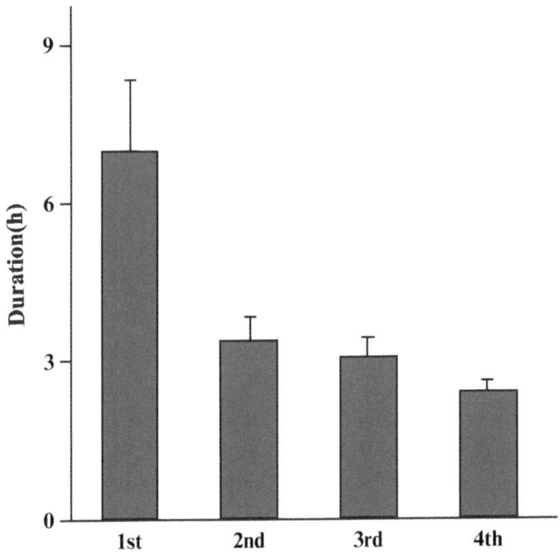

Fig. 4.8 The average time (h) from the beginning of nest foundation to the first defecation (*1st*) and the average intervals from one defecation to the next after the first defecation (*2nd*, *3rd*, *4th*) with SE bar (from Sato et al. 2003) in *St. miscanthi* females

From the biparental defense as well as from the cooperative nest building and utilization (nest sanitation) behavior, it can be concluded that *St. longus* and *St. miscanthi* have a highly developed level of sociality (now considered communal sociality; Mori and Saito 2005), the first such discovery in the Acari. Furthermore, Mori and Saito (2005) indicated that as nests become smaller, the counterattack behavior decreases (Fig. 3.24), meaning that a conflict exists between nest size and social development. As expected, large nests can contain many individuals and can persist longer as both food sources and living spaces (Sect. 3.5). Because the defensive success increases with adult density per nest (Fig. 3.22, Box 3.7), an additional strategy becomes apparent in *St. longus* and *St. miscanthi*, i.e., making larger nests to enhance the efficiency of cooperative nest defense by many members, even though these larger nests are more vulnerable to predator invasion. As such we see cooperative nest building, cooperative nest sanitation, and cooperative nest defense in the large nest building *Stigmaeopsis* species. Although cooperative feeding by adults is an important criterion for defining cooperative breeding (Wilson 1975), I have not yet observed such a behavior. One reason why there is no cooperative feeding might be related to phytophagy, because juveniles always live on a food source, i.e., host leaves, on which there is no difficulty in feeding by themselves. In other words, there may be no need for parents to feed their young (Box 4.6).

Meanwhile, I wish to discuss some by-product discoveries from our studies on fecal manipulation mechanisms. *St. miscanthi* lives on *Miscanthus* grass and *St. longus* on *Sasa*; both these species are large nest builders. Therefore, whether

Box 4.6

The possibility that there are other kinds of parental behavior in these mite species cannot be rejected, e.g., egg sanitation as protection from fungal disease, as known in earwigs (Lamb 1976). Mori et al. (1999b) observed that the eggs of *Stigmaeopsis longus* with a parent female showed a slightly higher survival rate than those without a parent female under field conditions, although the difference was not significant. Furthermore, Kikuchi (unpublished) discovered that a fungus species that is dangerous to *St. longus* sometimes occurs in its nests. A recent study suggested that parent female attendance is important to serve eggs within nests under field conditions (Kanazawa, unpublished data).

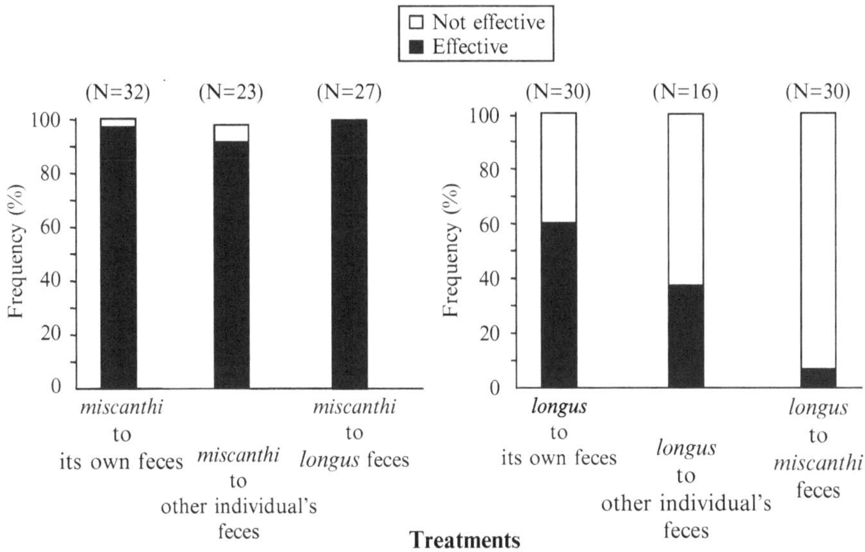

Fig. 4.9 Effectiveness of feces of different species as cues of waste management (from Sato et al. 2008)

these two species developed their communal societies independently becomes an interesting question. There are three kinds of evidence that suggest the importance of host plant shift in their speciation. The phylogenetic relationship hypothesized by ribosomal DNA (28 S) variation between the *Stigmaeopsis* species weakly suggests that *St. miscanthi* and *St. longus* might have originated from a common ancestor inhabiting *Sasa* bamboo and/or bamboo (Sakagami et al. 2009). From the fecal manipulation study, *St. miscanthi* can respond to the volatile chemicals of *St. longus* feces, but *St. longus* cannot respond to those of *St. miscanthi* feces (Fig. 4.9; Sato et al. 2008). Because we now know that the volatile chemicals originate from the host plants, the fact that *St. miscanthi* retains an ability to use chemical cues from *Sasa* bamboo suggests that *Sasa* is its ancestral host plant. The third piece of evidence discovered is that *St. miscanthi* immatures can develop on *Sasa*, but *St. longus* immatures cannot develop at all on *Miscanthus* (Sakagami et al. 2009). Taking these three observations together strongly suggests that *St. miscanthi* evolved from a common ancestor of *St. longus* living on *Sasa* bamboo and that the host plant shift from *Sasa* to *Miscanthus* might have played a great role in this speciation event (Sakagami et al. 2009). Therefore, we are now thinking that communal sociality evolved in *St. longus* (or its ancestor) once, then a social *St. miscanthi* branched out through host plant shift.

4.5 Sociality in Spider Mites

So far, I have only addressed sociality in the genus *Stigmaeopsis*. Are there any developments in sociality in other spider mite species? The answer may depend upon how sociality is defined, i.e., in the narrow or broad sense. As stated briefly when discussing life type variation, there is diversity in defecation behavior among spider mite species. In the WN life type especially, we now know that most species have more or less fixed defecation locations in their nests; e.g., in *Schizotetranychus brevisetosus*, *Eotetranychus suginamensis*, and *Eotetranychus shii*, nest members deposit their feces on the outside surface of the woven nest roof, *Eotetranychus querci* and *Schizotetranychus shizopus* place their feces around the inside margin of their nests, and *Schizotetranychus saitoi*, *Schizotetranychus approximatus*, and *Schizotetranychus laevidorsatus* defecate just outside of their nests (Box 4.7). Such life type features strongly suggest that this behavior is to prevent nesting spaces from becoming dirty. In their nests, eggs and immature stages are cohabiting, such that this behavior could be considered a kind of parental care. Furthermore, the nest-building behavior itself serves to protect offspring from predators, as shown in *Sz. recki* in Sect. 3.4. Therefore, we could say that species having WN life types possess parental care behavior, which has long been considered a prime criterion of subsociality (Wilson 1975; Saito 1995b). Waste management is not always restricted to the WN life type, however. As stated previously, several species having CW life types only deposit their feces on a CW web. This behavior must prevent the leaf surface from becoming dirty, but I hesitate to define such mites as subsocial based solely on this behavior.

In Figs. 2.3–2.6, I have shown the various social behaviors known in spider mites. In these figures, I adopted three criteria to define the level of mite social development as follows: the length of parent–offspring attendance, nest building by cohabiting parents, and offspring and feces management behavior, as considered in *Stigmaeopsis*. Readers of this book may differ in their views and some may perhaps think that there is no merit in defining such species as subsocial or primitively social. Even if so, I believe that close relationships (other than sexual ones) between individuals must give us some background into social evolution in animals. The first criterion is closely related to life history patterns in spider mites because parent–offspring attendance is primarily determined by whether parents have sufficient

Box 4.7

Why don't Thailand species, whose life types closely resemble Japanese *Stigmaeopsis* species, have the habit of depositing their feces at particular sites? My co-researcher, Y. Sato, had a fascinating suggestion, i.e., the flush toilet hypothesis. In tropical regions, very heavy rainfall frequently washes plant leaves. She thus supposed that there is no need for such species to manipulate their feces, because rainfall always removes any feces that have been deposited outside nests. Actually, these species deposit feces all around the perimeter of the nest under experimental conditions, but we could not observe any feces there in the fields. Of course, it is still just a story.

In relationship to the above, another curious finding is that *Schizotetranychus saitoi* (Thailand species) lives on the upper surface of bamboo leaves, although the other two species in Thailand and all species of *Stigmaeopsis* live on the undersurface. As described, although very heavy rainfall in tropical areas must make it more difficult for *Sz. saitoi* to live on upper surfaces of leaves, why has it such a habit? The fact that the nest made by this species is very solid seems to be related to this question, but I have no idea what factors have caused *Sz. saitoi* to live there.

time to attend their offspring. In Sect. 2.2, I addressed life history variation among several species (see Fig. 2.1). It is apparent from this figure that there is a great overlapping of generations (at most three generations) in *St. longus*. Mothers oviposit for a long time and the relatively short developmental time of immature stages creates generational overlapping. Such life history peculiarities thus provide a background for *Stigmaeopsis* sociality (a reminder that human beings also have such peculiarities). If we agree to this, all spider mite species enjoy more or less the same background, although there is variation in the extent of generational overlapping among species.

Aggregating at a certain place for a long period is also a prerequisite of social organization, because close interaction between individuals is only possible if they have frequent contact (viscosity). This requirement is realized by nesting behavior or chemical control (aggregation and dispersion pheromones are known in acarid mites; Kuwahara et al. 1982; Kuwahara 1990). In spider mites, there is little evidence of aggregation by chemical cues [although there are sex pheromones (Cone 1979) and attractive odor from the plants infested by conspecifics (Pallini et al. 1997)], but we do know that there are plant mite species that form large aggregations on the plant leaf surface without any nest-like structures (Box 4.8).

As such, in the narrow sense at least, subsociality is only recognized in spider mite species having WN life types. Of course, if we extend the definition of subsociality to include species showing short-term egg care, i.e., egg web covers, most spider mite species could be categorized as primitively social or subsocial.

Box 4.8

We have observed a tremendous aggregation of *Tydeus* sp. on the flat undersurface of durian leaves in Thailand (see below). We sometimes observed similar aggregations of tydeiid and asciid mites on plant leaves. Kuwahara et al. (1982) reported aggregation pheromones in acarid mites living on stored products, but there is no information about the functions of such aggregations on plant leaves. As stated in the text, aggregation is one of the prerequisites of social evolution, such that these are interesting from an evolutionary point of view.

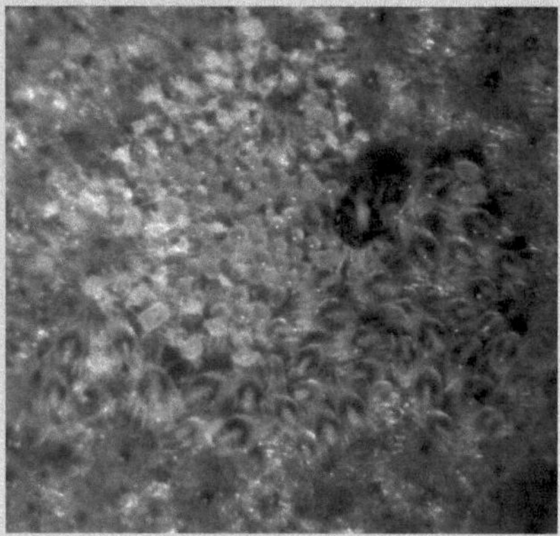

4.6 Overview of Mite Sociality

Are there any mite species other than spider mites that have social behavior? It has been reported that several mite species belonging to various families have sociality. Some predacious cheyletid mites belonging to the order Actinedida (spider mites belong to the same order) are supposed to show sociality because they live gregariously in nests built with silken threads. *Hemicheyletia (Hm.) morii* shows several special behaviors (Mori et al. 1999a). They use thin threads to construct nest-like structures that resemble woven mats at the curled edges of the undersides of leaves. The web-mats serve as footing to prevent individuals catching powerful prey from being dragged away from their nesting site. Eggs are oviposited on the web mats and then covered by anchoring threads (Fig. 4.10). A nest usually contains many individuals of various stages (Fig. 4.11), but there are no males, suggesting thelytokous reproduction (rearing under experimental conditions showed this species is thelytokous; Mori et al. 1999a). All active mite stages stand in a circle around the periphery of their woven mat, with their powerful pedipalpi opened toward the outside (see Fig. 4.10). When a prey item ventures too close and comes into contact with a cheyletid's pedipalpi, the cheyletid seizes a prey appendage (leg, pedipalpus, bristle, and so on) with its pedipalpi, then inserts its

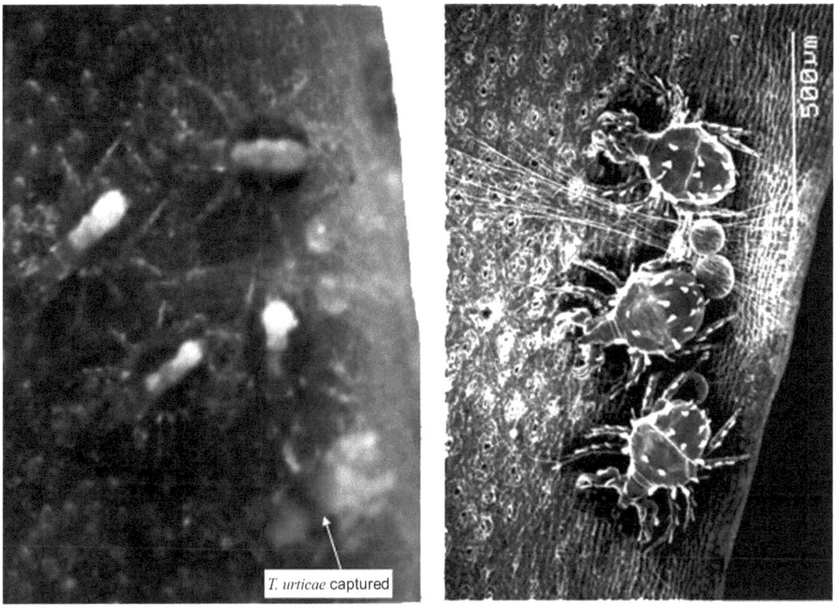

Fig. 4.10 Micrographs of *Hemicheyletia (Hm.) morii*. *Left*, using an optical microscope; *right*, SEM with cryo-system attached

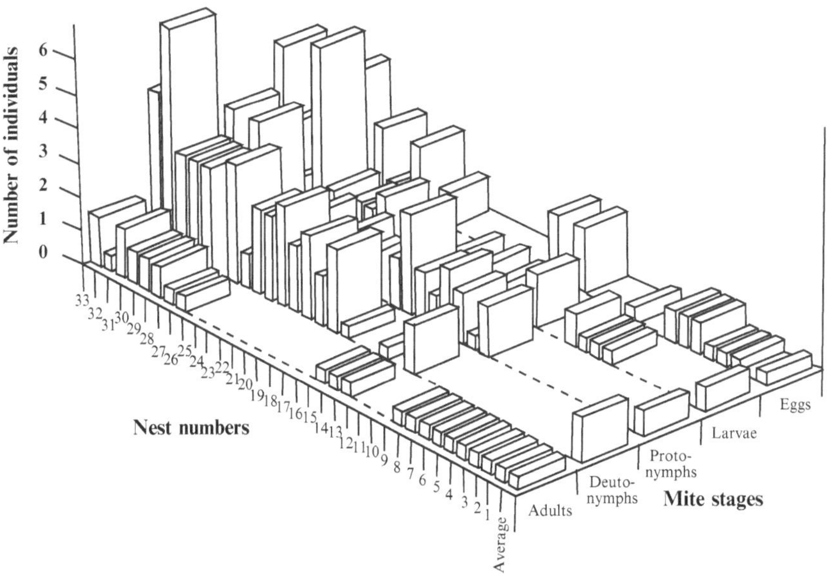

Fig. 4.11 Stage structure of *Hm. morii* nests (from Mori et al. 1999a)

chelicerae into the appendage (perhaps injecting toxin) and commences feeding. Immediately after a predator captures prey, other nest members swiftly respond to the prey's struggle by taking hold of other appendages and feeding on them. Through such group predation behavior, this mite species can capture very large, strong prey species relative to their body size, such as aphid larvae, thrips, and phytoseiid adults. There is no aggression between nest members even during feeding. This was the first discovery of a sit-and-wait group hunting species in Acari (Mori et al. 1999a; but also see Cloutier and Johnson 1993).

Cooperation between predacious conspecifics is not a common phenomenon in arthropods, although several examples have been reported in spiders and non-spider arachnids (Aviles 1997; Uetz and Hieber 1997; Rayor and Taylor 2006). The reason for such extraordinary hunting behavior, i.e., group predation in *Hm. morii*, can be explained by both their thelytokous reproduction, which inevitably increases the relatedness between nest members, and their small size, which must require cooperation to extend their menu of victims. We sometimes observed that if an *Hm. morii* individual captured a large prey item – an adult phytoseiid female, for example – without any help from other individuals, it was soon dragged away from its web mat, suggesting that the cooperation of several individuals is necessary to capture large and strong prey items. Furthermore, through this behavior, parent females can provide food to their offspring, because immature stages always get feeding opportunities when their parents (and other group members including themselves) capture prey. A marvelous behavior observed in this mite is that one of the nest members usually tries to throw the victim's dead body from their nest after predation, which may function as nest sanitation as well as cleaning the trapping site (Mori et al. 1999a). From the aspect of social organization, such a behavior may look like a kind of altruism for the other individuals, but I think that this is only under the rule that the last individual feeding the victim must do so even if it is alone. Anyhow, we should categorize this species as having cooperative sociality. In relationship to this, Boczek (1959) reported that the *Cheyletus eruditus* females protect their clutches until the larvae emerge, suggesting that there are other chyletid mite species that show a social behavior.

Another example appears in *Varroa jacobsoni* (Donzé and Guerin 1994). This species is a male-haploid mite parasitizing honeybee broods. The mites accumulate their feces at a particular site on the wall of the honeybee's brood cell and use it as a rendezvous site for mating. Furthermore, the foundress only feeds at a single site on the bee larva, probably increasing the survival of her offspring through the avoidance of drowning as a result of the hemorrhaging of the host.

There is also an interesting example of finely organized subsociality in moth ear mites. I believe that subsociality in the Acari must have originally been discovered in this group by Treat (1975), although he understatedly said that he had insufficient data. Several *Dicrocheles phalaenodectes* females, which parasitize the ear chambers (tympanic organ) of noctuid moths, utilize different chambers of the ear for different purposes: the first and second chambers are used for feeding and oviposition, and the third is used for mating and the deposition of feces. Thus there is a regulatory mechanism for public sanitation between room members. Furthermore, adult mites defend their chambers against intruders. Because the chambers include many offspring, this defense behavior may represent a kind of parental care, even if they do not give any special care directly to offspring. This species is assumed to be male-haploid (Norton et al. 1993).

Some phytoseiid mites aggregate and sibling individuals interact without cannibalism. In one interesting report of predatory behavior in phytoseiid species, Cloutier and Johnson (1993) observed that young nymphs of *Neoseiulus* (*Amblyseius*) *cucumeris* have difficulty preying on thrip larvae by themselves, such that they can develop normally only when conspecific gravid females hunt the prey for them. Although Cloutier and Johnson did not consider this phenomenon as social organization, I suggest that these mites perform a kind of parental care, i.e., feeding their young. As previously mentioned (Sect. 2.5), there is another kind of social organization, i.e., avoiding cannibalism in phytoseiids and *Hp. aculeifer* (Usher and Davis 1983). *Macrocheles superbus* shares oligochaete worms (Blazak et al. 1990), but Walter and Proctor (1999) said that cooperative hunting is unlikely, because the initial attack is by a single mite and then either the thrashing of the prey attracts others or they stumble across the dying prey and commence feeding.

Furthermore, I intend to address some fragmental knowledge about *Tp. bambusae* and *Ag. iburiensis*, both of which have been introduced as important predators of *Stigmaeopsis*. As I have described in Sect. 2.5, *Tp. bambusae* has nonfeeding larvae, which was hypothesized to be an adaptation for avoiding sib-cannibalism (Chittenden and Saito 2001), and females of this species seldom cannibalize their own offspring (Saito, unpublished data). When I first discovered the counterattacking behavior of *St. longus* (Saito 1986a,b), I simultaneously reported on the difference in counterattacking behaviors between *St. longus* females and males (see Sect. 4.4): When an adult female *Tp. bambusae* invades an *St. longus* nest where two females are cohabiting, one of the females could usually escape from the nest while the other female was being eaten (see Fig. 4.6). On the other hand, the predator female will always eat two male *St. longus* if they co-defend a nest (Fig. 4.6). From the point of view of *St. longus*, these phenomena can be easily explained as follows: *St. longus* females have another chance of reproduction elsewhere, but males do not, such that males are strongly concerned with the survival of their offspring even if they can rarely win against such a strong predator (Sect. 4.4). However, if we change our point of view from the prey to the predator, we may attain a different explanation: The predator female that intrudes a new prey nest first sweeps for dangerous males (and females) that will counterattack against her offspring (Saito 1986a). The observation that *Tp. bambusae* females seldom cannibalize their own offspring under no-food conditions now inclines me to believe the latter possibility.

Ag. iburiensis also shows a kind of social behavior. In Saito et al. (2008a), we discovered that the scattered nest effect hinders this predator (Sect. 3.5). At the same time, however, we observed that this predator female has a tendency to repeatedly return to nests that it has previously occupied (although this trend was statistically insignificant). Because there are usually several eggs laid by the predator female in such nests, this tendency may be related with maternal care, i.e., subsociality.

Although this habit seems to enhance the effect of void nests as decoys, there are undoubtedly reasons for the predator female to do this. It may not be so unusual as to think that the predators of social prey species could also be social. Of course, this is still only an idea.

Anyhow, with the exception of tetranychids, the level of social organization that suggests a beginning stage of sociality is only known in a small number of species scattered over several different taxa. An important point from such a rough review of sociality in mites is that all examples other than tetranychids are predators or animal parasites. I have no idea how to explain this. On the other hand, it has been suggested that gregarious living within narrow spaces, such as in bee brood cells, and the associated contact of individuals can result in close interactions between sibs. Fecal manipulation behavior is inevitable to realize long-term gregarious living in the same space; thus Donzé and Guerin (1994) suggested that this is a key behavior for the evolution of sociality. In the same context, Jackson and Hart (2009) recently discussed about the importance of the reducing mechanism (blind gut) of fecal load in eusocial evolution in hymenoptera. I do not think that the statement by Donzé and Guerin (1994) is always true, because such behavior is merely a prerequisite for gregarious living and/or only a result (not a cause) of social organization.

Last, I briefly consider the Rayor and Taylor (2006) review of sociality in non-spider arachnids. They examined the patterns of parental care, duration of association, and the presence of social traits found in the most social taxa of non-spider arachnids. Species in most arachnid orders have transient parental care with defense of eggs, a brief period of association with newly emerged young before independent foraging and explosive dispersal from the natal nest. More prolonged sociality, with long-term associations among mother–offspring–sibling, is rare and has only been described in a few species in the Amblypygi, Scorpionida, Pseudoscorpionida, and Acari. All such species have subsocial origins (i.e., there is no example of parasocial origin; cf. Wilson 1975).

4.7 Background of Social Evolution in Mites

There is no doubt that several factors or backgrounds are important for the evolution of sociality in mites. Some of these have been already mentioned. Furthermore, Tallamy and Wood (1986) stressed the peculiarities of resources, and Crespi (1994) suggested that one of the requisite conditions for the evolution of eusociality (caste) is possession of shelter-like structures providing food. Rather than repeat such comprehensive reasoning here, I would prefer instead to concentrate on factors peculiar or common to mites. Of course, the peculiarities mentioned previously, such as life history, genetic systems, sex ratio, and sexual behaviors are all fundamentally related to mite sociality (Box 4.9).

Hamilton (1964) first formalized that cooperative behavior, especially altruistic behavior, evolved under kin selection (inclusive fitness concept; Box 4.10):

Box 4.9

Phylogenic constraints, behavioral and ecological traits, and environmental factors related to the evolution of the sociality of several *Stigmaeopsis* species.

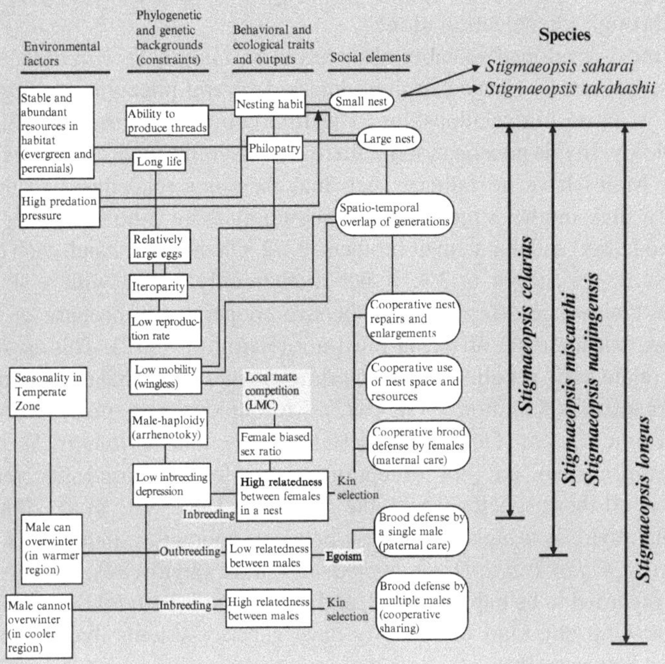

Box 4.10

The concept of inclusive fitness provided by Hamilton (1964) is a key to understand his kin selection theory. He defined two fitness components: one is personal (individual) fitness and the other is the fitness in other kin individuals; he formalized that gene evolution must occur through these two passageways. This is the famous kin selection theory: if your genes exist in your full sib-brothers and sib-sisters at the probability of 0.5, then your life is just equal to two brothers or sisters from the aspect of gene evolution.

If there is a sufficiently high relatedness between interacting individuals and the actor exhibiting the trait gives great benefits to recipients with low associated costs, such a cooperative trait will evolve under kin selection (through inclusive fitness = "personal fitness + fitness through relatives"). As previously stated, low mobility, overlapping generations (continuous iteroparity and rapid development), and the aggregative nature of kin create a high probability of sib interactions occurring. Therefore, it can be said that mite sociality evolved on the subsocial route (Wilson 1975), i.e., parent–offspring interaction, as stated by Rayor and Taylor (2006). If so, can we really say that the sociality of mites has evolved through kin selection alone?

In the most developed sociality observed in *St. longus*, *St. miscanthi*, and probably *St. nanjingensis*, I have certainly detected several phenomena suggesting the existence of close kin relationships. The first comes from their genetic system, male-haploidy. In this genetic system, there is an asymmetry in relatedness between offspring. Males have no fathers, such that the son's relatedness to their father (accurately, their mother's partners) is 0 (assuming there is no relatedness between mother and father) and that to their brothers is 1/2. On the other hand, each daughter's relatedness to her sisters is 3/4, if her mother only mated with a single male. Therefore, the female offspring must behave cooperatively because of their high relatedness, but the male offspring must not (Hamilton 1972). This asymmetry in offspring relatedness is believed to be a prime mover of eusocial evolution in haplodiploid organisms (Hamilton 1964, 1972). Most spider mite species have a haplodiploid genetic system (Gutierrez and Helle 1985), and *St. longus*, *St. miscanthi*, and *St. nanjingensis* are not exceptions; very female-biased sex ratios were observed in all these species (see Table 4.3; Saito 1987; Sato et al. 2000a; Zhang 2002). Thus there is a high relatedness between interacting individuals in a nest. *Hp. aculeifer*, *V. jacobsoni*, *D. phalaenodectes*, most phytoseiids, and most tetranychids are regarded to be haplo-diploid, and *Hm. morii* is thelytokous. Thus, all mite species showing any kind of sociality have genetic systems that guarantee high relatedness between sibs.

The second phenomenon concerns their life cycles and physiological adaptations: as shown in Fig. 4.12, most overwintered *St. longus* females establish their nest solitarily in spring, although two or more females (foundresses) were also observed (Saito 1987). Among the spring foundresses, there were unexpectedly high numbers of unfertilized females in both *St. longus* (8.7–21.1%; Saito 1987) and *St. miscanthi* (3.3–36.0%). If such unfertilized females found spring nests solitarily, they will only lay male offspring, because they have male-haploidy. In fact, the unfertilized *St. longus* females only lay a small number of haploid eggs (=male), then become inactive (similar to diapausing females; Saito 1987). Because male overwintering is extremely rare in cooler regions (males have no diapause adaptations; see Sect. 2.4), these phenomena strongly suggest that one trait of *St. longus* females when they are left in an unfertilized state is to wait for fertilization by their sons. Such mother–son mating is a kind of self-fertilization, and it greatly increases the relatedness between nest members in the successive generations (Box 4.11; Saito 1987). Furthermore, it is an important genetic peculiarity that factors causing inbreeding depression are generally scarce in haplo-diploid organisms. (I address this problem again in Chap. 5.)

Third, these mite species show high philopatry with low dispersion rates. Basically, *St. longus* males do not often disperse but mate with their nest mate

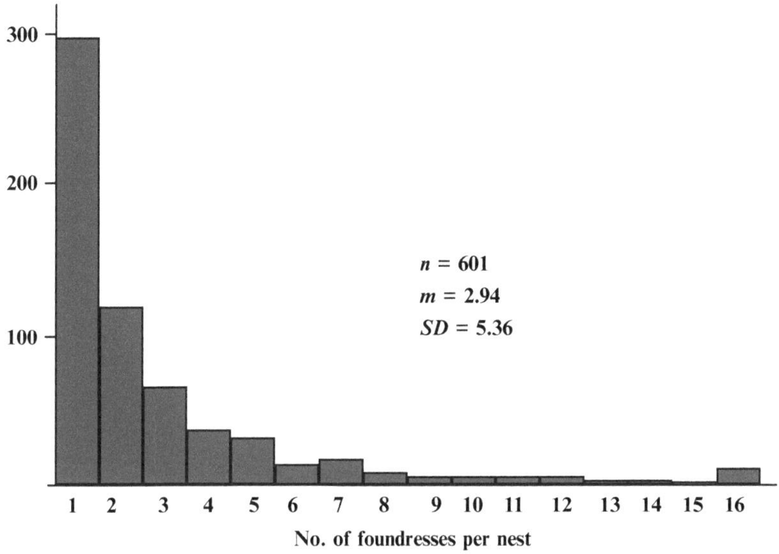

Fig. 4.12 Spring nest foundation pattern in *St. longus* (from Saito 1987)

Box 4.11

How the single foundation of an unfertilized female increases relatedness between male nest members in *Stigmaeopsis (St.) miscanthi*. No male immigration or male survival until the next generation was assumed.

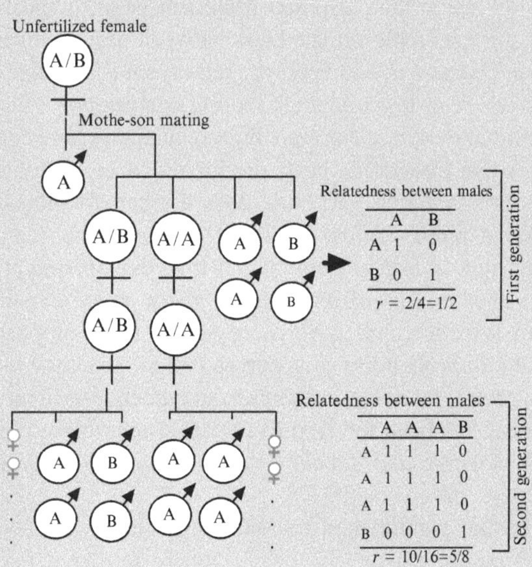

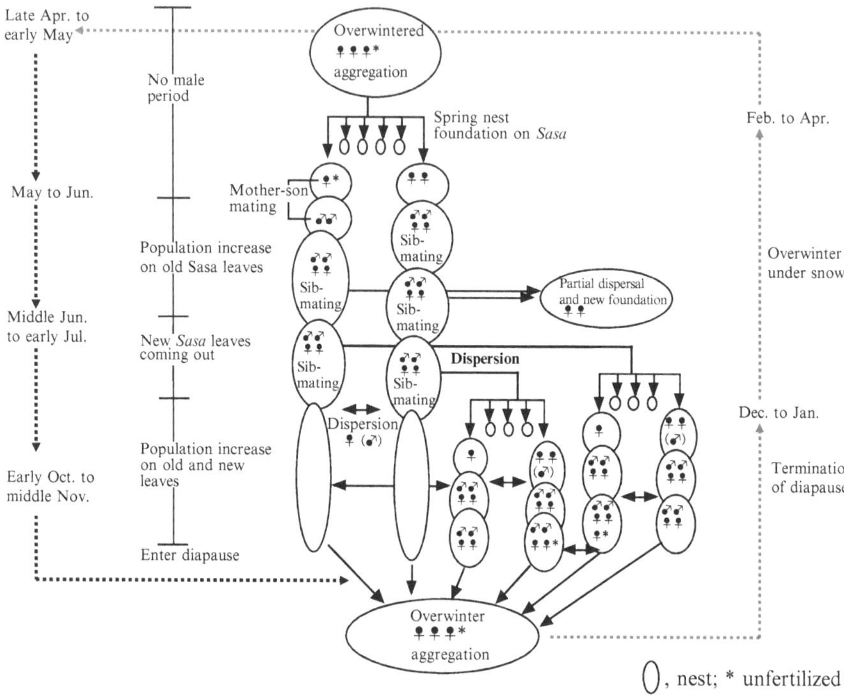

Fig. 4.13 Basic life cycle of *St. longus* in Sapporo

females. Some of these mated females then disperse to establish new nests, although most females reside inside their parental nests and begin oviposition (Fig. 4.13). Such a life cycle suggests a high relatedness between nest members. *V. jacobsoni*, *D. phalaenodectes*, *Hm. morii*, and WN type tetranychids are regarded to be phylopatric and are also more or less under sib mating conditions.

Such circumstantial evidence suggests high relatedness between the individuals of the mite species that live within nests or closed spaces, although no direct data about relatedness are available yet. How does the benefit increase by such kin-selected traits? What kinds of cost are there? Without such data, kin selection, despite its obvious appeal, remains only a plausible explanation at best.

There may be another line of explanation about social evolution in mites; it concerns the group selection model. Maynard Smith (1964) proposed the "haystack model" to show the improbability of group selection, at a time when the Wynne-Edwards (1962) group selection reasoning had been prevalent (see Box 4.4). Although this model has been referred to for explaining biased sex ratio in mites (see Sect. 4.3; Nagelkerke and Sabelis 1996), I once again address whether this model is applicable to mite sociality.

Let us consider that a number of haystacks (habitats with food) emerge at the beginning of an "animal season" in a field, where fertilized females can singly invade and reproduce for several generations (Fig. 4.4). All the haystacks are exploited for a certain period; thereafter, all individuals leave their haystacks simultaneously and gather in a particular place to mate (note that this situation is a little bit different from Sect. 4.3). The mated females again individually invade "haystacks" that have regenerated in the field. The necessary conditions of this model are (1) haystacks must, for a time, be reproductively isolated, (2) haystacks must be started by one foundress (randomly selected from all mated females), and (3) there is no dispersion between haystacks (the food supply is discontinuous in space; each patch of food supports its own haystack group). Let us denote that the probability of assortative mating is R (mating between the individuals from the same haystack) and that of random mating is $1-R$. The haystacks invaded by single "timid" females mating with "timid" males can produce "$1+G$" offspring, while those invaded by single "aggressive" females mating with "aggressive" or "timid" males only produce "1" offspring ("1" does not mean the real number of offspring here, but a unit, thus a relative value). Under such circumstances and life cycles, there are conditions under which genes coding for cooperative traits (timid–scramble) can increase their frequency by suppressing egoistic traits (aggression–contest) depending upon G and R values. Maynard Smith (1964) showed this with a one-locus and two-allele model. He showed that the condition necessary for the evolutionary spread of "timid" behavior is $R(1+G)>1$. If there is no inter(out)breeding between colonies, even at migration ($R\approx 1$), timid behavior will evolve, provided it is advantageous to the group; i.e., genes producing more offspring in a haystack increase. On the other hand, if all dispersers mate randomly, i.e., $R=0$, there is no condition where the "timid" trait will evolve. For an intermediate R value, there must be conditions under which the "timid" trait will evolve depending upon the G value, although Maynard Smith (1964) stated:

> If the admittedly severe conditions listed here are satisfied, then it is possible that behavior patterns should evolve leading individuals not to reproduce at times and in circumstances in which other members of the species are reproducing successfully. Whether this is regarded as an argument for or against the evolution of altruistic behavior by group selection will depend on a judgment of how often the necessary conditions are likely to be satisfied.

As described so far, the more or less inbreeding nature of spider mites and other social mite species, as well as their nesting habits, are thought to partly satisfy the conditions hypothesized in the haystack model: *St. longus* females tend to establish nests singly (Fig. 4.12), the nests persist long enough for the reproduction of several generations, offspring females mate with their sib males in their maternal nests before dispersal (it is analogous to high R), and their cooperation in nest defense may increase their offspring survival (higher G). Therefore, I have to agree with the possibility that the haystack model can also explain the social development of mites, although we have no accurate data whether the three necessary conditions of Maynard Smith are satisfied.

Rather, I think, however, that the kin selection and the haystack models are not alternative models, because habitat structure is not considered in the former and relatedness is not considered in the latter. If we calculate the relatedness between the members of the haystack in the latter model, we may immediately notice that most of them are very close kin. In other words, the habitat structure hypothesized in the haystack model inevitably creates a situation under which individuals having beneficial traits for kin interact strongly. Thus, these two forms of reasoning may be regarded as two sides of the same coin, I believe.

Chapter 5
Inbreeding Depression in Haplo-diploidy

As stated before, there is little doubt that kin selection is an important prime driver of social evolution. Inbreeding sometimes causes high relatedness between individuals, as is the case with social spider mites (note that inbreeding is a result of life patterns but is not a prerequisite for kin selection). However, in the animal kingdom, inbreeding is believed to be deleterious because of inbreeding depression (Charlesworth and Charlesworth 1987). There are two contradictory views on the genetics of haplo-diploidy in relation to inbreeding effects. One prediction is that there are few deleterious genes in haplo-diploidy because such genes had been selected out through haploid (hemizygotic) males (Smith and Shaw 1980; Atmer 1991). The other prediction is that there is a sufficient number of deleterious genes that only affect diploid females even in haplo-diploid organisms (Crozier 1985). I once contemplated the possibility that the genetic peculiarities of haplo-diploidy might be related to the fact that there are so many eusocial species in haplo-diploid organisms (Saito 1994b). We then tried to find out whether there are such genetic peculiarities in spider mites.

Crozier (1985) pointed out that spider mites are good model organisms for understanding the genetics of haplo-diploid organisms, because hymenopteran insects having haplo-diploidy and eusociality are difficult animals for genetic studies. Using a social spider mite, *St. miscanthi*, we observed how strong inbreeding, namely mother–son mating, causes inbreeding depression. As a result, we could observe significant inbreeding depression in female fertility with the increase of Wright's *f*-value (Wright 1922), but no depression in immature survival (Fig. 5.1; Saito et al. 2000b). Furthermore, we detected that the genes causing this depression effect are mildly deleterious recessive genes (Fig. 5.2b; Saito et al. 2000b). Through the inbreeding experiments, we could establish two lineages by mother–son mating for 11 generations; i.e., one lineage had no depression (normal lineage) and the other lineage showed about one-half decrease in fertility by strong depression (depressed lineage). Then we conducted a cross-experiment between them. As shown in Fig. 5.2a, the F_1 females produced by the cross between these two lineages oviposited at a similar rate to the normal lineage. Then we back-crossed the F_1 hybrids with normal and depressed lineages. The F_2 females obtained from the back-cross showed an intermediate oviposition rate between normal and depressed lineages

DOI 10.1007/978-4-431-99456-5_5,

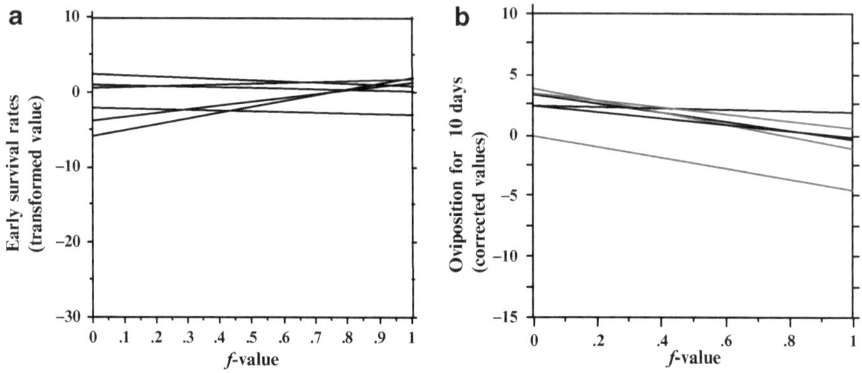

Fig. 5.1 Inbreeding effect on oviposition was very significant, but not on immature survival, in *Stigmaeopsis miscanthi* (from Saito et al. 2000b). **a** Indirect inbreeding effect on the survival rate of progeny females (produced by inbred pairs). **b** Parental inbreeding effect on the reproductivity of female progeny (observed). Number of eggs laid for 10 days was corrected by environmental variance observed in the outbreeding control. Thin lines in b show significantly depressed lineages.

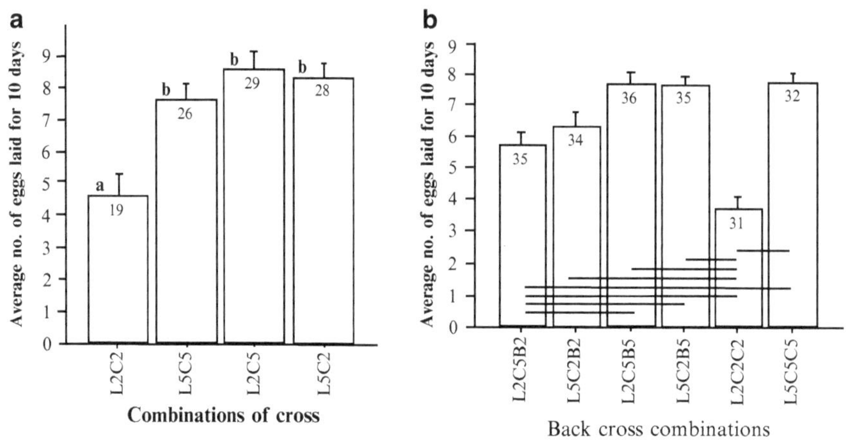

Fig. 5.2 Cross-experiments clearly showed that depression in female fecundity is caused by some deleterious recessive genes. *L2* is a lineage depressed by inbreeding; *L5* is a lineage not depressed by inbreeding in Fig. 5.1b. **a** Fecundity of fertilized females produced by parental crossing. *L2C2* offspring of L2-female×L2-male, *L5C5* those of L5-female×L5-male, *L2C5* those of L2-female×L5-male, *L5C2* those of L5-female×L2-male. Between different alphabet characters, differences are significant at $P<0.01$. **b** Results of back-cross experiments using the crossed lineages. Between the combinations linked by *horizontal bars*, there are significant differences at $P<0.03$. *Vertical bars* are standard deviations. (from Saito et al. 2000b)

(Fig. 5.2b). Therefore, we could determine that the genes governing the depression of oviposition are recessive.

The foregoing discovery (Saito et al. 2000b) is important, because it was the first paper in which clear inbreeding depression by recessive genes was detected in haplo-diploid organisms (see also Helle and Overmeer 1973). Thus we could prove Crozier's (1985) hypothesis that recessive deleterious genes exist mostly in the female. However, we met a difficulty in our attempts to publish this discovery, because we started the inbreeding experiment from parents that originated from two distant (~300 m) populations to gain genetic variability in the parental lineage. There is a difficulty in regarding this result as evidence of inbreeding depression, because of the possibility of heterosis in the parent generation. It was impossible to determine whether the depression was caused by the breakdown of heterosis or by the expression of deleterious genes retained in a population, because we had started our inbreeding experiment from a mixture of two adjacent populations. If we focus on a short-time process only, then this is a valid criticism, but we should expect such heterosis to readily occur among field populations in nature, because these sub-populations are adjacent. Anyhow, our primary purpose was to ascertain whether there are deleterious genes masked by dominance-recessive genetics in haplo-diploid females, and we presently believe that our findings have sufficiently proven this to be so.

However, the doubts still lingered. Then we (Mori et al. 2005) conducted a similar inbreeding experiment using parental lineages originating from a single population occurring far from the populations used in Saito et al. (2000b). There was very little depression on immature survival of inbred lineages in all populations. On the other hand, in the two of eight inbred lineages, all of which originated from small populations, female oviposition decreased significantly with the increase of Wright's f-value, showing that mildly deleterious genes are actually retained even in natural populations of haplo-diploid organisms (Box 5.1).

From the foregoing two studies (Saito et al. 2000b; Mori et al. 2005), we could demonstrate two important facts; i.e., deleterious genetic factors causing depression of female fecundity actually exist in haplo-diploid mite populations, and such depression is not always caused by the breakdown of artificially created heterosis, but by genetic factors retained at the semi-isolated natural population level. If the inbreeding depression observed in Saito et al. (2000b) was a chance result of an artificial crossing, we would have a very slim possibility of observing it again in

Box 5.1

Henter (2003) recently reviewed the nature of inbreeding depression in haplo-diploidy in comparison with diplo-diploidy and showed several important phenomena. She detected severe inbreeding depression in longevity that is not obviously sex limited in a haplo-diploid species. The discrepancy between her results and ours may be caused by differences in approach and materials. In our two studies (Saito et al. 2000b; Mori et al. 2005), inbreeding effects on several traits such as male insemination ability and adult male longevity were ignored, such that further studies on these traits are also necessary to understand in detail the genetic structure of haplo-diploid mites.

field populations. However, we could show that the factors causing depression are frequently retained in wild mite populations (Mori et al. 2005). This result means that effects such as the high frequency of outbreeding (it may sometimes cause heterosis), pleiotropic effects, and/or mutation–selection balance are operating in natural populations (Crozier 1985; Saito et al. 2000b).

As such, we know that most inbreeding effects in haplo-diploidy are female sex limited and that animals possessing this genetic system have a high tolerance to inbreeding, i.e., low depression in immature survival. These peculiarities might permit spider mite inbreeding, such that the high relatedness between interacting individuals through mother–son and sib mating may create the conditions under which kin selection frequently operates.

Meanwhile, another plausible factor for social evolution can be detected from the nature of inbreeding depression in haplo-diploidy. As observed by Saito et al. (2000b), reproductive inequality among the diploid females of haplo-diploid organisms is frequently caused by inbreeding depression. This means that we must observe "semi-sterile" females in haplo-diploid organisms more frequently than diplo-diploid organisms, because most females produced by inbreeding in the latter are selected out, especially as immature stages, by their low tolerance to inbreeding depression. Therefore, Saito (1994b) hypothesized that such differences in tolerance to inbreeding must be related to the frequent occurrence of nonreproductive castes in females of haplo-diploidy. For example, *St. miscanthi* females having deleterious homozygous recessive genes suffer depressed reproduction, but most can still construct their nest webs normally (Saito, unpublished data). If such web nests serve as protective refuges and females perform nest defense for their relatives (possibly including sisters having recessive and normal genes heterogeneously), we may simultaneously recognize "semi-sterile castes = depressed females" and "reproductive castes = normal females" within a nest. By this reasoning, I have proposed a possible genetic constraint that may explain some of the still-unresolved problems in eusocial evolution, i.e., a mechanism for the occurrence of reproductive castes, an explanation for why caste differentiation only occurs in females in haplo-diploidy, and for the frequent evolution of eusociality in haplo-diploidy. Whether such reasoning is flawed, as per Keller's (1995) objection, or plausible must be determined by future studies (Saito 1995c). It was recently shown in ants that interlineage crossing, thus outbreeding, causes caste differentiation (Glennis et al. 2002). This process is completely contradictory to my reasoning, because it results in reproductively suppressed castes (workers), which are heterozygotes. However, the important point is that both my own reasoning and the discovery in ants showed that eusocial evolution is sometimes caused by "genetic factors." In any event, I believe my reasoning may contribute to future hypotheses about the origins of eusociality in other haplo-diploid organisms (Saito 1997).

Chapter 6
Kin Selection

So far, I have discussed that mite sociality may evolve under kin selection, although there still remain several other possibilities. In this chapter, I address more detailed and more realistic phenomena on kin selection in mites.

6.1 Variation in Male Aggressiveness Between Species

While observing the social behavior of *Stigmaeopsis (St.) longus* and *St. miscanthi*, I noticed a considerable difference in male pugnacity between these two species. I subsequently observed male-to-male interactions in *St. longus* and *St. miscanthi* as follows. Web nests in which *St. longus* or *St. miscanthi* females deposited several eggs before the following experiments were provided. After all females had been removed from their nests, two strange *St. longus* or *St. miscanthi* males (from different stock cultures of the same species) were released there. These nests were then observed for 5 days. There was a considerable difference in male-to-male behavior between these two species (Fig. 6.1). *St. longus* males cohabited in nests for long periods and showed no aggression toward each other. On the other hand, *St. miscanthi* males always engaged in mortal combat that inevitably resulted in the death of one of the paired males during the 5-day period.

In the nests containing many individuals, multiple *St. longus* males show pre-copulatory mate guarding of the same female (Sect. 4.2) without any aggression whatsoever (Fig. 6.2), but *St. miscanthi* males fight to the death when they meet in the nests. Furthermore, the victor often sucked the loser's body contents, i.e., showed cannibalism (Fig. 6.2). Since these initial findings, I have carefully observed many *St. miscanthi* nests in stock cultures and found numerous dried-up dead bodies of males inside and outside the nests. I am convinced that such strong killing behavior is ordinal in this species.

I was really surprised to observe such extraordinary male behavior in *St. miscanthi* because spider mites are herbivorous. Just what is it that *St. miscanthi* males are competing for through mortal combat? I supposed that the aggressive behavior of *St. miscanthi* males must be to acquire females in a nest, i.e., to form a harem.

Y. Saito, *Plant Mites and Sociality: Diversity and Evolution*
DOI 10.1007/978-4-431-99456-5_6, © Springer 2010

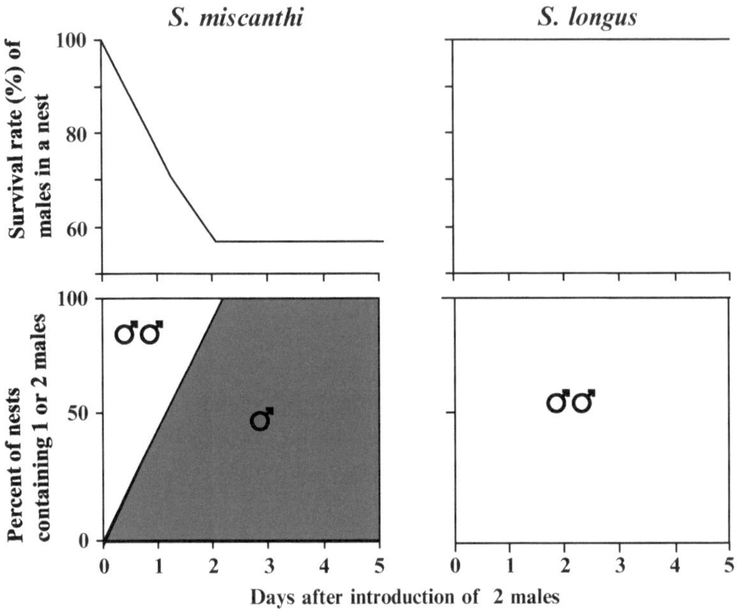

Fig. 6.1 There is a major difference in male pugnacity between *St. (S.) miscanthi* and *St. (S.) longus* when two males are introduced into a nest (after Saito 1990c; Saito and Mori 2005)

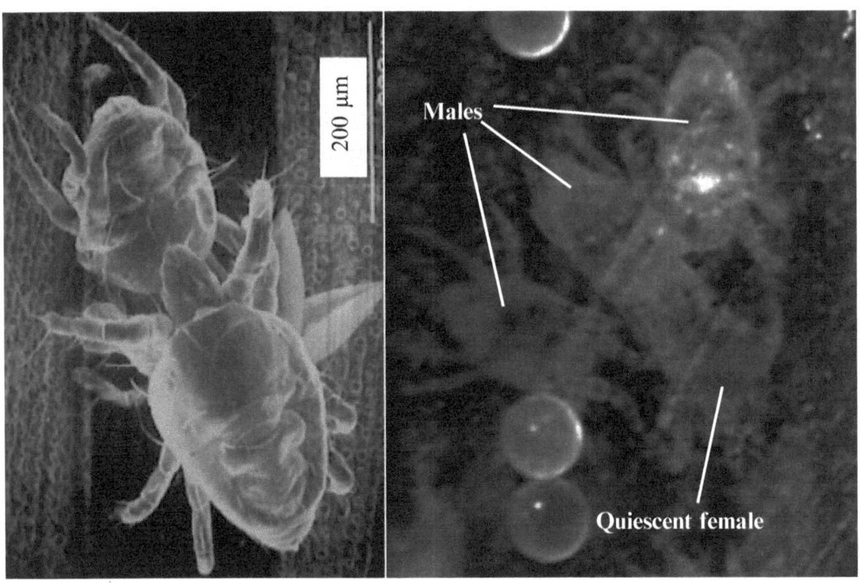

Fig. 6.2 Multiple male precopulatory guarding in *St. longus* (*right*); *St. miscanthi* male killing conspecific male (*left*)

To ascertain whether such a difference in male pugnacity between species really exists in nature, Saito (1990c) also observed the male distribution patterns of *St. longus* and *St. miscanthi* in the field, and revealed that *St. miscanthi* males establish their harems by killing and/or driving away all other males in a nest.

These findings provided me with another important theme about the evolution of "social and/or intrasexual traits." As is known in many animals, spider mite males are generally aggressive against conspecific males and sometimes kill each other to acquire mate priority (Sect. 4.2). Such male aggressiveness has been empirically observed in many species of Tetranychinae. Therefore, in contrast to my first impressions about the mortal combat of *St. miscanthi* males, the important point is actually the male amicability in *St. longus*. Evolutionary biology theory tells us that male aggressiveness inevitably evolved by intrasexual selection through egoism (Wilson 1975). Thus, I have challenged the reason(s) why there is such a difference in male-to-male behavior between *St. longus* and *St. miscanthi* (Saito 1990c,d).

The first study was concerned with the mechanisms that determine the winner male in combat in *St. miscanthi* (Saito 1990d). I hypothesized three factors – priority of nest, male age, and leg length – as possible determinants of the victor. Potter et al. (1976) already reported that leg length (as a reflection of body size) is one of the factors determining victory in *Tt. urticae* male-to-male combat. Then we observed paired male behavior in a nest in the same manner as before. Table 6.1 shows that there was no difference in winning probability between different-aged males or in the order of nest-occupying male (owner or intruder). On the other hand, males having relatively long leg length (leg I) showed significantly higher winning probabilities (Fig. 6.3; note that male leg length is only measurable once the mite has been prepared as a slide specimen) (see Saito 1990d). This result is the same as observed by Potter et al. (1976). The last result strongly suggests that leg I is a weapon in *St. miscanthi* male-to-male combat and that such a character must be under intrasexual selection (Saito 1995a; we return to this later).

The exceptionally aggressive nature of *St. miscanthi* males leads us to question whether such behavior only occurs between strange males or is also common between kin males. I next observed the relationship between fighting intensity and kinship (Saito 1994a). As stated earlier, this is a haplo-diploid species; unmated

Table 6.1 Owner/intruder and age effects on winner male in male-to-male combat in *St. miscanthi* (from Saito 1990d)

Situation (age in days)			Owner (or older) Victory	Owner (or older) Loss	Probability (two-sided)
	Age is				
A: Owner (5.3)	=	Intruder (5.3)	12	8	$P=0.54$
B: Owner (8.1)	>	Intruder (3.7)	11	6	$P=0.33$
C: Owner (3.8)	<	Intruder (7.3)	8	7	$P=1.00$
Owner (6.6)	vs.	Intruder (4.6)	23*	14*	$P=0.19$
Older (7.7)	vs.	Younger (3.7)	18**	14**	$P=0.60$

*The data of A and B were lumped by difference in ownership
**The data of B and C were lumped by difference in age

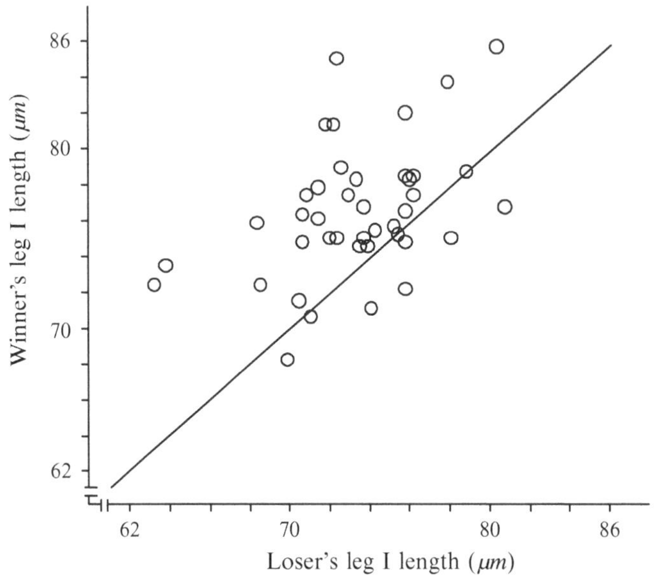

Fig. 6.3 Difference in size of leg I between winners and losers in male-to-male combat of *St. miscanthi* (from Saito 1990d)

Table 6.2 Death probability and time until death on one of paired males in kin-paired and stranger-paired male-to-male combat of *Stigmaeopsis (St.) miscanthi* (from Saito 1994a)

Experiments	No. of tests	Five days after pairing		Time until one male's death (mean ± SD days)
		Both alive (%)	One killed (%)	
Between brothers	53	9(17)	44(83)	2.8 ± 2.0
Between non-kin males	52	13(25)	39(75)	2.9 ± 2.9

Difference in death probability between two experiments was insignificant (Fisher's exact probability test, $P = 0.44$) and that in mean length of the time until a death was also insignificant (t test, $P = 0.93$)

females can produce only males among which there is at least 1/2 relatedness (of course, if the mother was also produced by inbreeding, then the relatedness between them is much higher). Thus, kin males were obtained from unfertilized females. Non-kin males were taken from different stock cultures, which had been established from at least 500 individuals collected in the field (Ashizuri population). The relatedness between the non-kin males could not be strictly estimated, but it was considered to be much lower than that between kin males. Then two kin males or two non-kin males were paired and released into a woven nest containing several eggs. During the 5-day observation period, there was no difference in male mortality rates between kin pairs and non-kin pairs (Table 6.2). In both cases, males fought strongly with and killed cohabiting males. Although *St. miscanthi* males do not seem to exhibit any kin-discriminative behavior, whether they have kin recognition ability is still unknown because we did not conduct any experiments to prove it.

As such, I learned that *St. miscanthi* males show very high aggressiveness against conspecific males regardless of kinship. At first I thought that this aggression was a kind of shift from "counterattack against predators" (Sect. 3.5) to attack against "conspecific males." In other words, the aggressive traits males exhibit against conspecific males are the same that they exhibit against predators. However, this kind of impression must be rejected because *St. longus* males also show very strong counterattack against predators yet show no aggressive behavior against conspecific males (see Fig. 6.1). Therefore, we have to separate the male aggression against predators from that against conspecific males. So, what are the possible reasons for the variation in male aggressiveness seen between *St. longus* and *St. miscanthi*? Saito (1990c,d) could only suggest that such a difference in male aggression may be related to a difference in male relatedness at the population level between these two species, because *St. longus* is mainly distributed in cooler regions and *St. miscanthi* in warmer regions in Japan, which may cause lower dispersal (or mixed colonies) in the former and higher in the latter.

6.2 Variation in Male Aggressiveness Within a Species

After completing the foregoing studies, I had an unlikely idea: If there is such variation in male aggressiveness between closely related species, then could male aggressiveness vary within species? Fortunately, *St. miscanthi* has a very wide distribution range, from the northernmost point of Honshu Isl. (lat. 26°6′) to Ryukyu Isl. (lat. 41°30′), because its host plant, Chinese silvergrass (*Miscanthus sinensis*), occurs throughout Japan (and in many other Asian countries, as well). Thus I started to consider whether there is variation in male pugnacity between populations of a single species (Saito 1995a). *St. miscanthi* was collected from various regions around Japan (Box 6.1), and many strains were established in our laboratory. As described, it is very easy to evaluate paired male mortality in a nest, such that I regarded the death rate of one of the paired males as "male relative aggressiveness" and evaluated each population by the same-garden analysis (populations originated from various regions are tested under the same experimental conditions). Very large variation in male aggressiveness among populations was detected; e.g., one of the paired males was killed immediately in almost all nests in the Ashizuri population, but no males were killed over 5 days in the Mashiko population (Saito 1995a).

What kind of environmental factors were responsible for this variation was the next theme. To search for factors, we need a working hypothesis. One plausible factor was differences in the relatedness of interacting males among populations. Hamilton (1979) observed interspecies variation in male aggressiveness between species whose wingless males were expected to have different kinship from the difference in female oviposition pattern into figs. He explained the variation by the following three hypotheses: The reasons why there is low aggression between close kin males in a fig must be *(1) many of the rival males are brothers, and a male doesn't care so much whether he or his brother does the mating; (2) many of the*

Box 6.1

Localities where *Stigmaeopsis (S.) miscanthi* were collected in Japan. *Circles* indicate where populations in Japan were observed by Saito (1995a). *Hatched circles* show populations whose males were demonstrated to be HG; *open circles* were LW. *Squares* indicate points where *S. miscanthi* was confirmed to exist but where aggressiveness has not yet been observed. *Filled circles and squares* (all must be LW) indicate heavy snowfall regions. (from Saito and Sahara 1999) HG and LW, see Sect. 6.4.

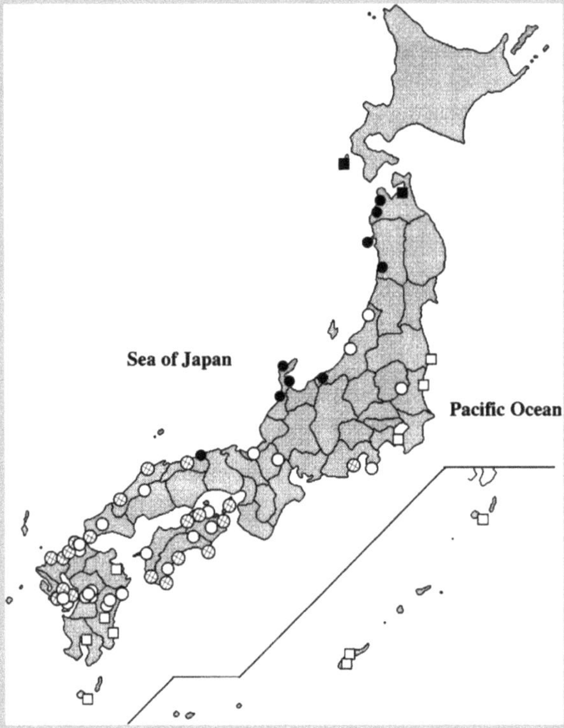

females are his sisters, and he doesn't wish to risk that some sisters remain unmated; and (3) for those that are his sisters, a male may actually prefer to have them mated by a unrelated male because of the opportunities this will give for useful recombination in the next generation. These points are all from his kin selection (inclusive fitness) theory. On the other hand, West et al. (2001) showed that there is no correlation between male aggressiveness and relatedness between interacting males per fig and the aggressiveness is rather correlated with number of females per male. They provided the "resource (mates) competition hypothesis" to explain the variation in male aggressiveness of fig wasps. We initially attempted to analyze the variation in accordance with Hamilton's (1979) kin selection theory.

From the kin selection hypothesis, the variation in male aggressiveness must be correlated with the relatedness of interacting males. If so, what factors can change the relatedness of males between populations? As stated before, I knew that the *St. longus* males distributed in cooler regions have no aggressiveness and those of *St. miscanthi* in hotter regions show high aggression, such that one of the candidate factors was temperature. I then attempted to analyze the correlation between the male relative aggressiveness and temperature of each sampling location. However, there are several kinds of temperature values, i.e., annual mean temperature, mean maximum daily temperature, mean minimum daily temperature, and so on. Among these, I selected mean temperature for summer (June to September) and the mean lowest daily temperature for winter (December to February), because mite life history, especially the number of generations (it must reflect the intensity of mixing by dispersal among populations), may mostly depend upon the former and male overwintering probability may depend upon the latter. The reason why I focused on the mean lowest daily temperature for winter should be explained further: *St. miscanthi* overwinters at the adult female stage (see Table 3.3) and the overwintering females enter diapause (tolerant to winter hardiness) in autumn (Saito et al. 2002). On the other hand, it has been pointed out that spider mite males have no such adaptations (for *Tt. urticae*; Veerman 1985). Although there are few adult males in spring (overwintered) populations of *St. miscanthi* in cooler regions (there are no males in overwintered *St. longus* populations in Sapporo; Saito 1987), some non-diapausing males sometimes overwinter in warmer regions in Japan (Saito 1995a). This finding must mean that the lowest temperature in warmer regions is sufficiently high to permit non-diapausing males to survive during winter (how many males can overwinter varies year to year depending upon winter severity). Male existence in spring may greatly influence the kin structure of this mite species, if unfertilized (unmated) females overwinter (Saito 1995a; Saito and Sahara 1999). As shown in Figs. 4.12 and 6.4, the foundation of spring nests is mostly carried out by a small number of females, although sometimes many females will build a nest. Therefore, if single unfertilized females found nests, they often mate with their sons (mother–son mating, as described in Sect. 4.7). Because such strong inbreeding has no short-term fatal effects for this mite species (Chap. 5), the relatedness between offspring produced by mother–son mating must increase in such nests (cf. Box 4.11). On the other hand, if males can overwinter, they must fertilize overwintering females in the winter aggregations (*St. miscanthi* and *St. longus* females usually form great masses in certain nests in winter; Saito, unpublished data). Thus, the probability of male overwintering (or survival during winter) is thought to affect the probability of inbreeding (mostly through mother–son mating; Saito 1987) in spring nests (Fig. 6.4). Therefore, winter minimum temperature, which may determine overwintering male survival, would reflect the relatedness among nest members in nests.

Figure 6.5 shows that the relative male aggressiveness is significantly correlated with the minimum temperature of the regions where mite populations were collected, although the mean temperature in summer did not show such a clear trend. If the minimum temperature in winter indirectly affects the relatedness of interacting

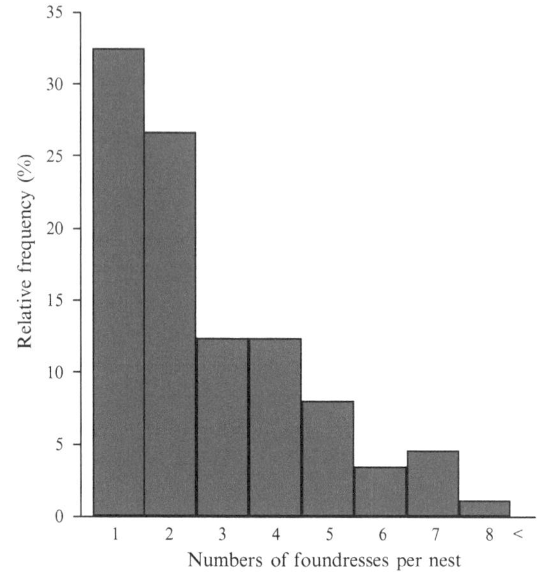

Fig. 6.4 Numbers of foundresses observed in spring (middle May) nests of *St. miscanthi*. Data from six populations from north to south Kyushu Island. Only nests with adults, eggs, and larvae were included, because only these could be regarded as newly founded nests

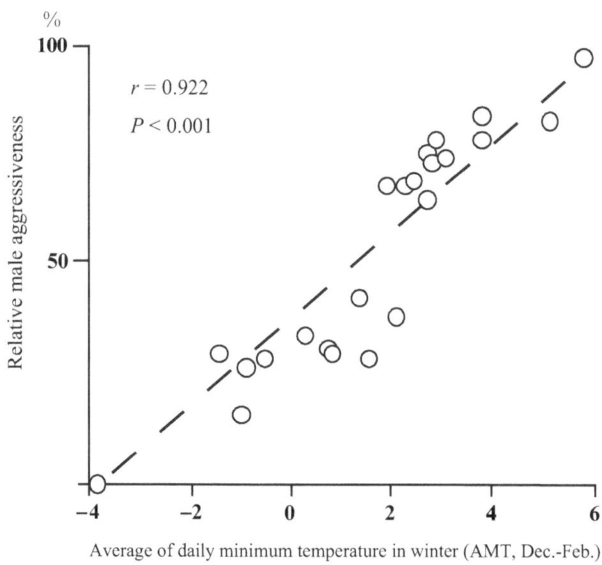

Fig. 6.5 Correlation between relative male aggressiveness and winter coldness. Relative male aggressiveness was estimated as the relative death rate of one of the paired males introduced into a nest for 5 days (after Saito 1995a). Each experiment included more than 25 replicates and controls

males, this correlation must suggest that male aggressiveness evolved by kin selection (Saito 1995a).

Next, let us verify whether the other hypothesis, "resource competition," is applicable to the present case. After publishing (Saito 1995a) the kin selection hypothesis, I met with several counter-arguments. Among these, the most popular opposition was as follows: a higher male mortality for whatever reason should bias the sex ratio. If so, the value of females as resources to males would decrease, and this might result in selection against male aggression. In other words, if females are easy to get, there is no need to fight for males, i.e., "resource competition" (West et al. 2001; see earlier). Then, I examined (Saito 2000; Saito et al. 2000a) whether this hypothesis holds in the case of *St. miscanthi.*

For the convenience of discussion, I divided the life cycle of *St. miscanthi* into two seasons: winter to early spring (the overwintering season) when they must overcome adversity, and summer when mites actively reproduce and increase their numbers. From winter to spring, male overwintering probability is low in cooler regions and high in warmer ones (Saito 1995a); this inevitably causes a difference in male abundance in winter to early spring. Under the kin selection hypothesis, this difference causes a difference in the relatedness of nest mates in the spring and summer seasons, whereas under the resource competition hypothesis the difference in male abundance directly alters the aggression tactics of males. As stated previously, male overwintering probability is lower in cooler regions than in warmer regions. With males abundant in warmer regions, we could expect higher aggression there in accordance with the resource competition hypothesis.

However, other difficulties associated with the resource competition hypothesis become apparent if we focus on the detailed population structure during the overwintering season. Even in relatively high aggression populations (such as the Tsuyazaki and Keya districts), overwintered males are much more scarce than females: males made up only 1.3% and 1.7%. respectively, of all adults observed in March (Saito 1995a). Therefore, I find it difficult to believe that such a small number of males, compared to females (resources), would display mortal aggression toward each other when competing for mating opportunities in this season. In other words, I believe that there are very few situations under which competition between males occurs during the overwintering season (including early spring) in *St. miscanthi*. On the other hand, as a rule of population genetics, a small number of immigrant males can greatly affect the genetic structure of a gene pool (Ims and Yoccoz 1997). Thus, male overwintering probability (male immigration for space is analogous to male survival over time in this situation) greatly affects the genetic structure, i.e., the relatedness of the colonies of this species (see also Sect. 6.4).

In the summer season, we studied the sex ratio at maturity of six populations of *St. miscanthi* (two of the "low-aggression" groups and four of the "high-aggression" groups mentioned hereafter) under experimental conditions and found very similar sex ratios among them; i.e., approximately 15% males (sex ratio of offspring produced by isolated female; Sato et al. 2000). This finding means that there is no difference in the basic sex ratio trait between populations. Moreover, Saito (2000) determined that there is no significant correlation between (experimentally evaluated) male aggression and the number of adult females per nest for 32 field populations. Rather, male aggression was negatively correlated with the number of males per nest in the fields. To avoid any confusion between the cause and consequence, namely, that mortal aggression of males decreases the number of males or vice versa, I investigated the relationship between a potential female sex ratio = (third quiescent females)/(third quiescent males + third quiescent females) and male aggression among populations ("third quiescent" is the last quiescent stage before maturation). The results showed that there is no significant correlation between them (Kendall's tau = 0.09, $P > 0.96$), indicating that the abundance of available (potential) females for potential males is not related to the variation in male aggression even in the summer season. Therefore, I could conclude that there is no strong evidence supporting the "resource competition hypothesis" (Saito 2000; Saito and Mori 2005).

6.3 Egoism and Cooperation

As a consequence, we have at least reached a conclusion that male aggressiveness in *St. miscanthi* must be linked to the relatedness between interacting males. However, the more serious question as to why males should decrease their aggressiveness when interacting with kin demands an answer. Even if frequent inbreeding increases the relatedness of nest members, males should still compete for mates. It is simply not rational that high relatedness between males must decrease male pugnacity. If there is even a small difference in mating success between males through aggression, sexual selection will favor more aggressive males (Sect. 4.2; Saito and Mori 2005). Thus, aggression might evolve in males of every population regardless of their relatedness when the relatedness, $r < 1$.

Saito (1995a) discussed this problem. Let us not forget the fact that *St. miscanthi* and *St. longus* males have a strong ability to counterattack against predators (see Box 3.7, Sect. 3.5) and that such counterattack behavior must become more effective when multiple adults defend a nest (see Fig. 3.22; Saito, unpublished data; and cf. Appendix 3). Under conditions of varying relatedness, *St. miscanthi* males must ultimately have two different interests in their nests: to improve their individual fitness through the monopolization of females, i.e., egoism (=direct fitness), and to achieve inclusive fitness by cooperatively defending their relatives against predators, i.e., kin selection. Males do not distinguish their kin from non-kin (Sect. 6.1; Saito 1994a), such that conflicts between males may basically be influenced by the "mean relatedness" of interacting males in respective populations. Thus, a male living within a nest with his "kin males" in cooler regions gains little advantage from mortal combat. Therefore, he rather tends to improve his inclusive fitness (via his kin) through cooperative defense. If, however, he often encounters non-kin males who, in warmer regions, occasionally intrude from other nests, he would retain a mating advantage by eliminating his competitors through fierce fighting. Therefore, kin selection would be a primary factor in variation in male antagonism, although in a different way from Hamilton's (1979) reasoning.

If so, then the reader may wonder why *St. miscanthi* males do not show kin discriminative behavior (as shown in Sect. 6.1; Saito 1994a). If they had this ability, then males would behave amicably toward their kin males and aggressively against strangers. I have no clear idea of how to definitively answer this question, but I can suggest that it is related to the phylopatric nature of this organism; i.e., they do not encounter males having very different relatedness within their nests frequently enough for such an ability to evolve (Saito 1994a), such that they do not use such precise information.

There is also the question of whether the variation in aggression seen in *St. miscanthi* is affected by the variation of predation pressure (e.g., S. Aoki, personal communication; Saito 1995a). If there were big differences in predation pressure between local populations of *St. miscanthi*, then males should decrease their male-to-male competition, because they should defend their offspring in a cooperative way. In other words, if it has killed all the conspecific males within its nest, the fitness of the victorious male can completely disappear because it becomes unable to effectively defend its nest, such that the intensity of predation is related to the variation. There are actually several co-occurring predator species in the habitats of *St. longus* and *St. miscanthi* (Saito 1990b; Chittenden and Saito 2001; Mori and Saito 2004), and although these predator species are hypothesized to be one of the driving forces behind speciation in *Stigmaeopsis* (Sect. 3.5, Mori and Saito 2004), we have no strong evidence that there is a large difference in predator fauna among populations. However, if such an effect is important, there is no doubt that kin selection, i.e., relatedness, is also responsible for male cooperation, because high relatedness always makes it easy for interacting males to cooperate (see Sect. 6.6).

6.4 Two Groups Showing Different Levels of Male Aggressiveness

As such, the variation in male aggressiveness of *St. miscanthi* was explained by kin selection. However, we later noticed that there were two distinct trends in Fig. 6.5; most plots of the more aggressive populations were positioned on the upper side of the regression line and those of less aggressive populations on the lower side of the regression line. If I had erroneously observed two groups having different male aggressiveness, all the foregoing scenarios would be meaningless. Therefore, I was forced to study many more populations from various districts.

With the steady accumulation of data, it became apparent that there are two groups showing different levels of male aggression in Japan. As seen in Fig. 6.6, at an approximately intermediate value of relative male aggressiveness, there are two distinct groups, both of which show clinal variation (hereafter the low aggression group is called the LW form and the high aggression group is the HG form). Furthermore, with two parameters, i.e., relative aggressiveness and winter minimum temperature, a cluster analysis (Ward method) revealed that there are two forms separated by a long distance (Saito and Sahara 1999). In relation to this,

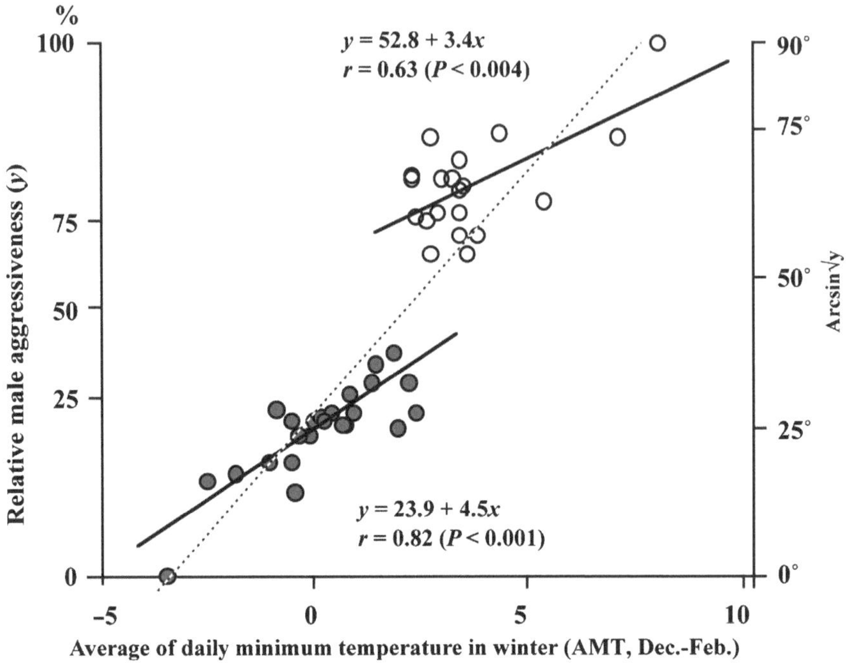

Fig. 6.6 Relative male aggressiveness (y) and average daily minimum temperature in winter (x) in populations from low-snowfall regions. *Open circles* are HG (high aggression) form; *filled circles* (*grey*) are LW (low aggression) form (after Saito and Sahara 1999)

Osakabe et al. (1993) had analyzed protein differences by two-dimensional electrophoresis in five populations of *St. miscanthi* and recognized two distinct groups: two populations in warmer regions (Okinawa and Kagoshima) and three in cooler ones (Mashiko, Yokosuka, and Wakamiya). Although these five populations do not always correspond to the populations observed in Saito and Sahara (1999) (only Mashiko and Wakamiya were included in the less aggressive populations of this study), the distribution pattern suggests that the former two belong to HG found in warmer regions and the latter three to LW found in cooler regions. Therefore, we may conclude that there are two subspecies (or forms), both having similar clinal trends in the relationship between male aggressiveness and relatedness created by winter coldness. In other words, male-to-male antagonism varies even within different "subspecies" in an equal manner and suggests that such a phenomenon is not always rare in nature. If so, what are the factors dividing *St. miscanthi* into these two subspecies or forms? We return to this later.

In addition to these results, we could simultaneously discover another piece of evidence that indirectly supports the kin selection hypothesis, i.e., "the winter coldness" and "male overwintering probability" relationship. In Japan, the amount of snow accumulation varies considerably among cooler regions; e.g., heavy snowfall on the west coasts (facing the Sea of Japan) yet very little snow on the southeast coasts of Honshu Isl. (facing the Pacific Ocean, Box 6.1). Since Kojima et al. (1970) reported that the temperature under snow varies according to depth and approaches 0°C at ground level, it was expected that the temperature affecting male overwintering survival is not the ambient temperature (i.e., from meteorological data), but the under-snow temperature in the regions of heavy snowfall. Therefore, it was expected that the relationship between male aggressiveness and average minimum winter (air) temperature varies between areas of light and heavy snowfall. Under this supposition, Saito and Sahara (1999) compared the relationship between male aggression and winter minimum temperature in two categories, light snowfall and heavy snowfall regions. As a result, male relative aggressiveness really depended upon the coldness of light snowfall regions (lower cline in Fig. 6.6), but there was no correlation between aggressiveness and temperature in heavy snowfall regions (Fig. 6.7). Therefore, the correlation between male aggressiveness and male overwintering probability must change according to the insulating effect of snow cover. The present comparison between heavy and light snowfall regions may demonstrate that male overwintering probability is not affected by the ambient temperature conditions of localities but by the temperature of the plant leaf surface where males are living. Simultaneously, this connection means that "male survival in winter" is a key to explaining male aggressiveness. Although it remains indirect evidence, it does lend support to the hypothesis proposed by Saito (1995a) that male aggressiveness (=intensity of intrasexual selection) evolved from a change in relatedness among interactors (=advantage of male cooperation).

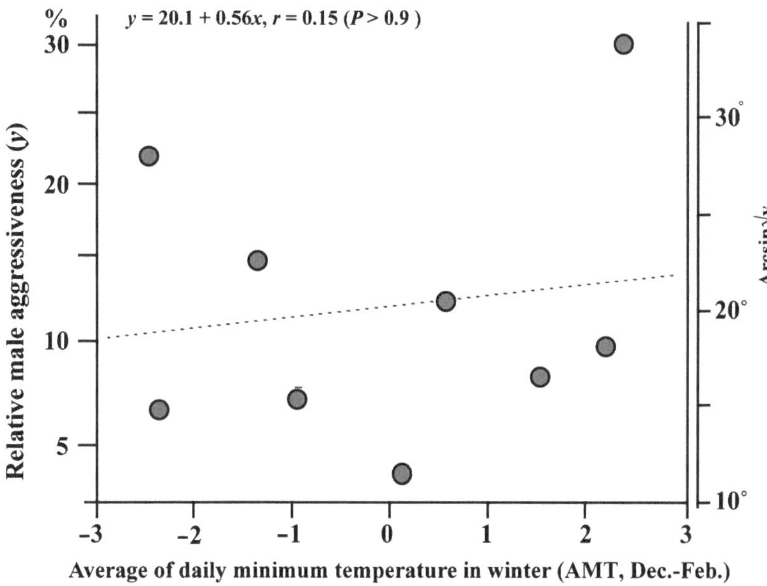

Fig. 6.7 Relationship between relative male aggressiveness and average daily minimum temperature in winter in heavy snowfall regions (after Saito and Sahara 1999)

Why Are There Two Groups Showing Different Clinal Trends?

The following may be the final question concerning the variation in male aggressiveness, but it has not been sufficiently resolved. Saito et al. (2002) observed the diapause attributes of the two forms (HG and LW) of *St. miscanthi*. We started the study under the working hypothesis that female diapause attributes also affect the relatedness between nest members in spring. For example, we have sometimes observed all stages of *St. miscanthi* (HG) on Chinese silvergrass in low-altitude southern regions even during the coldest season (Fig. 6.8 left). This observation suggests that some HG populations do not enter diapause but continue to reproduce during winter. Under such conditions, strong inbreeding (mother–son mating), which is thought to be the main factor increasing relatedness, must not occur. On the other hand, there are only females in winter in northern and/or high-altitude populations (Fig. 6.8 right), suggesting they enter deep diapause. Although the reason is still unknown, if such a difference in female diapause is a qualitative one, i.e., there are diapause and non-diapause populations, we should be able to explain the existence of the two groups showing different clinal trends. This idea came from several previous discoveries about diapausing and non-diapausing populations of *Tt. urticae* and *Pn. citri*

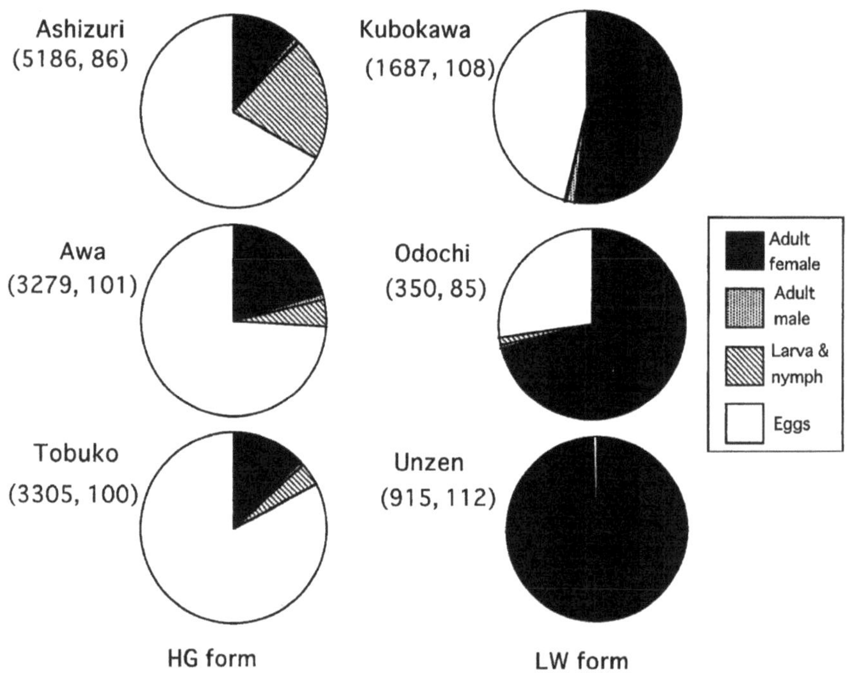

Fig. 6.8 Stage structure of six populations of *St. miscanthi* in midwinter. *Numerals in parentheses* are the number of individuals (*left*) and the number of nests (*right*) observed. *HG*, high aggression form, *LW*, low aggression form (from Saito et al. 2002)

(Shinkaji 1979; Gotoh and Shinkaji 1981; Osakabe and Saito 1991). At the present time, *Pn. citri* is divided into two species, *Pn. citri* (non-diapause) and *Panonychus mori* (diapause) by morphological differences (Osakabe and Sakagami 1994; Ehara and Gotoh 1992). The status of *Tt. urticae* is still disputed (Dosse 1967; Takafuji et al. 1991).

We then conducted an experimental study on the diapause attributes of the LW and HG forms (Saito et al. 2002). Two populations of each form were selected, and the diapause induction and termination patterns of females were then observed. Generally, spider mite diapause is induced under short-daylength and low-temperature conditions (Veerman 1985). First, we reared eggs of each population to adulthood under these conditions. After female maturity and mating, they were kept under different daylength conditions and their oviposition was observed. Figure 6.9 shows that the females of all populations developing under short-day and low-temperature conditions significantly decreased their reproductivity after moving to long-daylength conditions, indicating they enter diapause (Fig. 6.9b,d). Furthermore, there was a big difference in diapause elimination patterns between the LW and HG populations (Fig. 6.9), suggesting that there is a distinct difference in female diapause attributes.

However, data from only two populations of each form cannot provide sufficient information to draw conclusions. We next compared the stage structure of three populations of each form in the field (Fig. 6.8) and the winter temperature data of

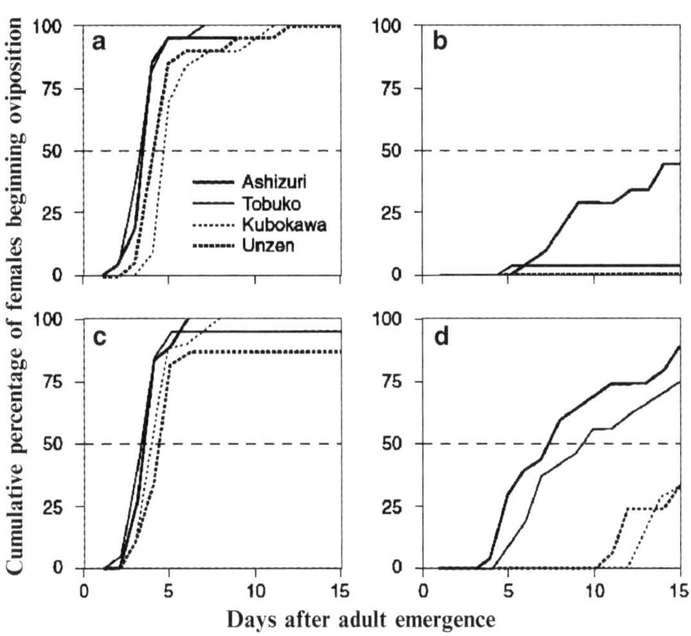

Fig. 6.9 Oviposition (ovi.) patterns at 18°C of females developing (dev.) under 18°C and under long-daylength and short-daylength conditions. **a**, development under 15L-9D/oviposition under 15L-9D; **b**, dev. 9L-15D/ovi. 9L-15D; **c**, dev. 15L-9D/ovi. 9L15D; **d**, dev. 9L-15D/ovi. 15L-9D Ashizuri and Tobuko are *St. miscanthi* HG form and Kubokawa and Unzen are LW form (from Saito et al. 2002)

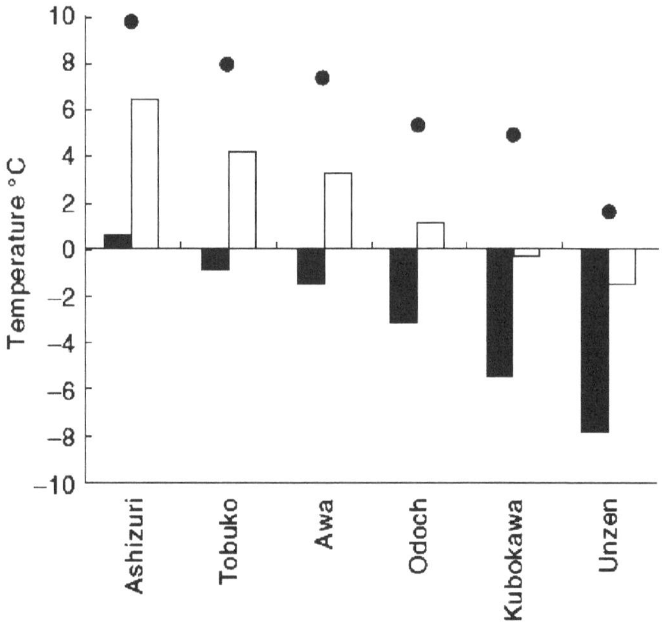

Fig. 6.10 Average winter (December–February) temperatures of six regions where *St. miscanthi* populations were observed. *Black circles*, average daily mean temperatures from December to February; *white bars*, 3-month (Dec.–Feb.) average daily minimum temperatures; *black bars*, monthly minimum temperatures for 10 years (calculated from AMEDAS, data base provided by the Japan Meteorological Agency during 1985–1995) (from Saito et al. 2002)

their respective localities (Fig. 6.10). Many eggs and developmental stages were found in the Ashizuri, Awa, and Tobuko populations (all HG), whereas the Kubokawa, Odochi, and Unzen populations (all LW) consisted mostly of adult females and a small number of eggs (Fig. 6.8). There were significant differences in stage structure among all populations. On the other hand, the three kinds of temperature data in the six localities where *St. miscanthi* were collected change more or less in a gradual manner (Fig. 6.10), such that it is difficult to understand why the stage structures of the HG form (Ashizuri, Tobuko, and Awa populations) are very different from those of the LW form (Unzen, Kubokawa, and Odochi populations). In addition, the stage structure of the Awa population is very different from that of Odochi, although the temperatures are not so different. This finding means that winter climate does not fully explain the variation in *St. miscanthi* stage structure in winter and also suggests that there may be some qualitative difference in diapause intensity and/or in physiological response (e.g., the threshold temperature for oviposition) between them (we have continued this study for many LW and HG populations, and this difference has become increasingly evident; Saito et al., unpublished data). Thus, the working hypothesis that the difference in female diapause attributes causes disruptive change in the mean relatedness of nest members became increasingly probable.

As such, we could explain, in a somewhat strained manner admittedly, the disruption of two clinal trends in the relationship between male aggressiveness and relatedness in Japan. This explanation is, however only applicable if we a priori agree with the existence of qualitative differences in diapause attributes between HG and LW. This inevitably leads to yet another question: why there are two forms having different diapause attributes? The road to a comprehensive overall explanation is a long one, and regrettably I cannot progress any further at this stage because of a lack of data. However, I engage in some plausible reasoning in Chap. 8.

6.5 Sexual Selection

An overview of sexual selection in mites has already been presented in Sect. 4.2. In this section, I digress a little to describe a case study on this problem in *St. miscanthi*. I address whether the variation in male antagonistic behavior has an influence on male weaponry through intrasexual selection. Sexual selection theory predicts that male body size and/or weaponry must evolve, as known in horned beetles, deer, and elephant seals, if males fight strongly to acquire mates (Wilson 1975; Alcock 1979). As already mentioned, Saito (1990d) observed that leg length is one important factor determining the winner of male-to-male combat in a population of this species. As there is variation in male aggressiveness, it is expected that there is some variation in leg length between populations of *St. miscanthi*, especially between LW and HG forms. Saito (1995a) measured male leg length to address this prediction. Leg lengths vary between populations for several factors, such as host plant conditions (Saito 1990d), but the leg length ratio (I/III) does not change (Nakano, unpublished data). The softness of spider mite bodies makes it impossible to measure their body sizes. Thus, I adopted the ratio of leg-I length / leg-III length (the relative length of leg-I), because behavioral observations indicated that the male leg-III is only used for supporting the body and not for aggressive behavior (although whether leg-III is a true reflection of body size is unknown).

Figure 6.11 shows the correlation between male relative aggressiveness and relative length of male leg-I. There is a significant difference in the relative length of leg-I between HG and LW, with the former being more exaggerated. However, I could not observe any clinal trend in leg length along with aggressiveness within either form. As stated previously, there are significant correlations between the aggressiveness and the winter minimum temperature within both LW and HG forms, such that this result is not of constant concern to me. This is one of the important questions left for future study. Anyhow, the difference in the relative leg-I length between LW and HG has become a good criterion to morphologically discriminate these forms in successive studies (e.g., Sato et al. 2008).

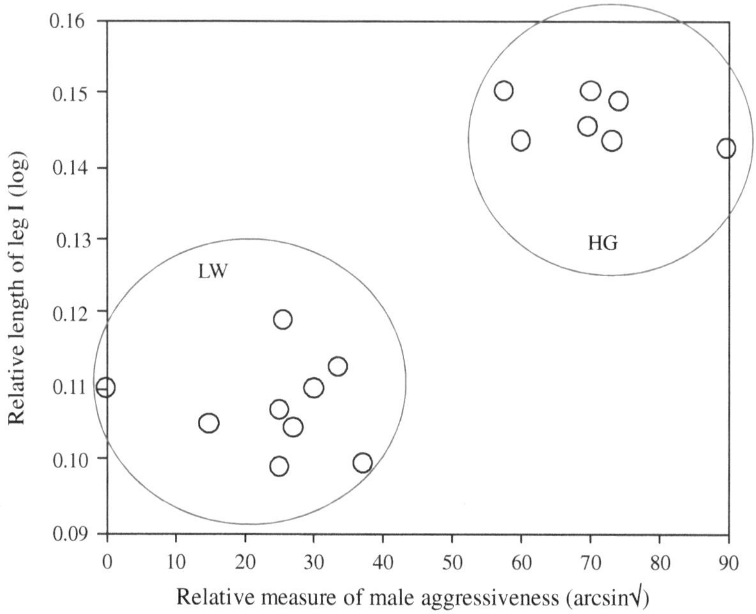

Fig. 6.11 Relative length of male leg I was not correlated with the intensity of male aggression (after Saito 1995a)

6.6 Theoretical Explanation of Male Aggression and Cooperation

Earlier I have described cooperation and aggression in males of *St. miscanthi* and *St. longus*. From the aspect of animal behavior, we have learned that there are two extremes, namely, aggression and cooperation in these two closely related species. Although aggression and cooperation, as fundamental themes in sociobiology, have always attracted a great deal of attention and research (Wilson 1975; Alcock 1979; Maynard Smith 1982a; Trivers 1985; Krebs and Davis 1987; Dugatkin 1997), they have always been treated as "separate" phenomena. For example, Dugatkin (1997) discussed animal cooperation through the game theory approach, but the strategies he hypothesized are a dichotomy of "neutral" and "cooperation." The famous hawk–dove game provided by Maynard Smith (1982a) and Hines and Maynard Smith (1979) also focused on two strategies, i.e., "aggression = hawk" and "neutral = dove." There are several reasons why these two behaviors have been treated separately. One must come from the fact that the prime movers (selection pressures) that evolved these two behaviors are considered to be different: aggression will evolve through egoism (individual selection), and cooperation through kin selection and/or group (or multilevel in the modern day) selection (Hamilton 1964; Maynard Smith 1982a; Dugatkin 1997; Lehman and Keller 2006). Furthermore, doubts remain over how an increase in inclusive fitness through cooperation between relatives (kin selection) influences individual fitness through resource competition between them (individual selection), and where (and when) these two extremes reach a compromise in group-living animals (Maynard Smith 1982b; Queller 1985; West et al. 2001; Giron et al. 2004; Saito and Mori 2005).

Absolute Conditions for Cooperation

Saito and Mori (2005) reviewed several hypotheses about variation in male aggressiveness and discussed how the diverse nature of aggression and cooperation can be understood if we focus on where aggression reaches a compromise with nonaggression and/or cooperation in response to the relatedness of the interactors. To further clarify this discussion, I regard aggression and cooperation as a "continuous trait" in group-living animals whenever possible. So, by considering relatedness, just how aggression reaches a compromise is addressed here. In the beginning, the outline of a famous and inclusive theory of aggression and/or cooperation must be redescribed, although I have previously introduced its essence (Sect. 6.2).

Hamilton (1979) stated, with respect to the variation in aggression of wingless fig wasp males, "a difference in mean relatedness between rivals accounts for the different male behaviours." Although he did not say that this is the sole factor accounting for the difference in behaviors (see Sect. 6.2), the above statement is sometimes believed to correspond to Hamilton's rule of altruism (Michod and Hamilton 1980), i.e., $rB-C>0$, where r is Wright's coefficient of relatedness to the recipient, B (≥ 0 per the primary definition) is the benefit associated with the trait for which the gene codes, and C (≥ 0) is the donor's cost, which accrues from the decrease in mating opportunities. This rule was then applied to explain the case of the male fig wasps (Trivers 1985; Frank 1985; West et al. 2001). However, as mentioned before, Saito (2000) doubted the rationale that high relatedness between males must decrease male pugnacity. If there is even a small difference in mating success between males through aggression, sexual selection will favor more aggressive males; i.e., aggression might evolve in males of every population regardless of their relatedness when $r<1$.

To ascertain the foregoing prediction, let us see whether "$rB-C$" tends to increase above 0 as r increases in male-to-male competition. The cost ($C\geq 0$) accrues from the withdrawal of an actor male (donor) from the competition for females and the benefit accrues from the increased mating chances for the other males (recipients) resulting from the actor's withdrawal. Assuming that females can be inseminated at all, let us imagine that there is a single male and n females in an arena and mating occurs between them. If x males join the arena, then the first (actor) male should compete with them. Are there any conditions under which the first actor male raises its inclusive fitness by sharing females? When the actor male decreases its aggression level and loses mating chances for d (=C), then the other males x will get a surplus benefit, d/x, each (assuming that the males joining the arena have primarily no chance of getting partners). Because $B=x(d/x)=d$, the inequality $rB-C>0$ is $d(r-1)>0$. Because $d>0$, $r>1$ is always required. In other words, the cost d paid by the first actor male as decreased mating opportunities is "intact" and is evenly divided among the other males as their benefits, so that there is "no way" of increasing the inclusive fitness of the first actor male under $r \leq 1$. From the beginning, therefore, it is unlikely that Hamilton's rule can ever be applied to the varying competition in mating between males, if both the competition and mating opportunity are not repeated (i.e., once in a lifetime; Maynard Smith 1982a).

Of course, the conditions under which fig wasp males compete are different from the foregoing example because they inevitably join multimale competition within a fig fruit. It is thus suggestive that the level of fighting between fig wasp males shows no correlation with the estimated relatedness of interacting males but is negatively correlated with mating opportunities (West et al. 2001). If we still intend to apply Hamilton's rule to male-to-male competition, we have to search for plausible conditions where any decrease in an actor's mating opportunity can significantly increase its kin's mating chances (i.e., $B \gg C$). I know that the above calculation is so simple that most people might consider it a self-evident conclusion. However, it is undeniable that some researchers believe that Hamilton's rule (note that this rule is only a part of his kin selection theory) is applicable to such a case. This problem is thought to be caused by an erroneous application of the Hamilton's rule: his rule is a comparison of fitness between solitary-living and group-living (interacting) individuals and shows the conditions under which the latter individual gets more inclusive fitness than the former one. In other words, it is not the rule to explain behavioral traits (aggression, cooperation, and so on) of individuals when they are always situated in game conditions (interaction is inevitable).

Several other hypotheses have also been put forward to explain variation in male aggression. One is the increase in lifetime mating success in nonfighting males (Enquist and Leimar 1990) or the high risk of disability to the winner male (Maynard Smith 1976; Tomkins et al. 2004; addressed later). Another is that if males are confronted with another different selection pressure (e.g., the synergistic effect of multiple males in Maynard Smith 1982b), then the selection pressure will also change male aggressiveness.

For the moment, however, let us examine whether such a different selection pressure can improve Hamilton's rule for explaining male aggression (i.e., by comparison between individual fitness and inclusive fitness). Let us recall *St. longus* and *St. miscanthi*. The males of both species are characterized by their counterattack behavior against nest-intruding predators (Saito 1986a, 1990e; Sect. 4.4). If there are multiple males in a nest, they may lose mating opportunities even if they effectively defend their offspring and/or mates. Therefore, males must inevitably adopt either of two traits for two different selections, namely, (1) cooperation in offspring defense against predators and (2) aggression to obtain mates. As described in Sect. 6.2, the variation in the aggressiveness of *S. miscanthi* males cannot be explained by the resource (=mate) competition hypothesis proven in the fig wasp case by West et al. (2001) but is thought to be closely related to the effects of the high risk of disability to the winner male and of cooperative defense by males. We first tried to formalize factor (1) as follows.

If a single male mates with females and defends his (female) partners in a nest, he will have f offspring over his lifetime. Thus, f is considered the net income (fitness) of a solitary male after his reproductive and defensive behavior. Cooperation of x males increases the survival of their offspring "$1+s$" times ($s \geq 0$). The reason why I express the fitness gain through cooperation as $1+s$ is to maintain consistency with the game model mentioned later. If the cooperators are his kin, he increases his inclusive fitness through the cooperators' offspring at the rate of r (relatedness). If x males live altogether, mating opportunities decrease and thus the number of offspring decreases because of the competition. Any decrease in mating opportunity for a male can be converted into the number of offspring "produced" by him, such that the cost is regarded as the decrease in the male's offspring to apply the same term as the benefit of cooperation. The fitness of a male belonging to an x-male group is

$$\frac{(1+s)f}{x} + r\frac{(1+s)f(x-1)}{x}.$$

Then we can search for the condition under which cooperation will evolve, i.e., "cooperating male's inclusive fitness" > "solitary male's fitness." Thus, we can obtain the following inequality:

$$\frac{(1+s)f + rf(1+s)(x-1)}{x} > f.$$

Because $f > 0$, it can be reduced to

$$(1 + s)(1 - r) + x(r + rs - 1) > 0.$$

If only two individuals interact ($x=2$), then the above inequality is simply expressed as $(1+s)(1+r) > 2$. This inequality can be further simplified as $r > (1 - s)/(1 + s)$ because $1+s>0$. The last formula suggests that if $s \geq 1$, cooperation will evolve regardless of the r-value, meaning that there is a condition under which cooperation occurs unless the interacting individuals are non-kin (e.g., Fowler and Gobbi 1988; Uetz and Hieber 1997; Matsuura et al. 2002). Furthermore, $r > (1 - s)/(1 + s)$ clearly indicates that male relatedness is also important; namely, r is the primary determinant of the evolution of cooperation, if $0 < s < 1$. Therefore, the effect of cooperation (s) and relatedness (r) must be essential to understand the cooperation, and here we once again encounter Hamilton's inclusive fitness concept (Hamilton 1964).

Next, let us see whether $r > (1 - s)/(1 + s)$ is applicable to the spider mite case. The parameter $s \approx 1$ (a twofold defensive success) evaluated experimentally in the case of *St. longus* (see Fig. 3.22) is now meaningful in accordance with this inequality. If $s > 1$, there is no option for the male mites other than cooperation regardless of r, while Saito (1990c, 1997, 2000) stressed the importance of higher r. Although the relatedness (r) between males may vary in these mite species, males have low dispersal trends, such that there is a high probability of interaction between close relatives. Therefore, both higher s and r can now explain why we only observed cooperative males in this species.

The foregoing explanation (Saito and Mori 2005) showed that aggression would be replaced by cooperation in relationship to the relatedness of interacting individuals at $r > (1 - s)/(1 + s)$. However, it does not show whether there are conditions under which aggression (or cooperation) gradually change (as seen in Fig. 6.5) with relatedness, the effect of cooperation, or other factors. Furthermore, the formula shows only the threshold at which "solitary living" changes to cooperation. In other words, it is still insufficient to resolve the question of whether the clinal change in male aggression observed by Saito (1995a) and Saito and Sahara (1999) can be logically supported.

Mixed Strategy of Aggression and Cooperation

I have stated that there is a difficulty in understanding variation in male-to-male aggression by Hamilton's rule. The situation is that the rule is a comparison of fitness between "solitary-living" and "group-living (interacting)" individuals and is not intended to explain behavioral traits (aggression, cooperation, and so on) of individuals when they are always situated under "game between more than two individuals" conditions. Animals that join "interaction" of conspecifics do not always choose such situations but inevitably join because of their life styles. If so, they have no way to choose "solitary living" under such situations (remember that males cannot obtain any fitness if they withdraw from mating competition!). Then, Saito and Takada (2009) solved this problem using a game theory model.

To ascertain whether there is a comprehensive explanation for the clinal variation of aggressiveness in *St. miscanthi* males, we examined a 2×2 nonrepeated game (thus without learning). The situation and content of the payoff matrix are as follows:

1. A symmetrical and nonrepeated game without learning occurs between two males in a patch (e.g., nest). All males in a population play the paired game.
2. Only two strategies, the aggression (abbreviated as A, hereafter) strategy and the cooperative helping (abbreviated as C, hereafter) strategy exist.
3. A and C strategies are not fixed for any male, so that each male plays the A strategy at frequency q and the C strategy at $1-q$. Any males adopting an A strategy in a game are called A strategists and those that adopt a C strategy are called C strategists for simplification.
4. Fitness is estimated by how many females the two male players can inseminate, which is equal to "how many offspring they can raise in the next generation." The net fitness of a single male without social interaction is assumed to be f as previously defined (note that this value disappears when all males play paired games).
5. We introduced two parameters, i.e., the cost of aggression, k, and the benefit of helping (cooperation), s.

The basic payoff matrix is then obtained as follows:

	C	A
C	$\frac{(1+s)f}{2}$	0
A	f	$\frac{(1-k)f}{2}$.

When a C strategist interacts with another C strategist, the net fitness is an additive benefit, and s ($s \geq 0$, hereafter called the helping effect) is shared equally between the two C strategists who defend their families; i.e., the payoff is represented by $[(1 + s)f]/2$. We have defined this situation as "cooperative sharing." If two males interact together in a patch (nest) with a constant number of females (mating partners), as known in *St. miscanthi*, one of the behavioral traits of such males is to equally share one-half of the partners (1/2). In cases where males possess the ability to defend and protect their patches (including their partners and offspring) from predators, then $s > 0$, as shown in Sect. 3.5. This helping effect (s) can be measured practically by comparing the success of defending offspring from predators by two C strategists with that of a single C strategist (an example is shown in Appendix 3).

If a C strategist interacts with an A strategist, it always loses its direct fitness and receives nothing (payoff = 0; i.e., no offspring). Although this situation may seem quite severe for the C strategist, the mortal male-to-male aggression in *St. miscanthi* (Fig. 6.1; Saito 1995a) indicates its reality.

If an A strategist interacts with a C strategist, it receives a payoff, f, because it always wins and can monopolize all females.

When an A strategist confronts another A strategist, they always fight. The winning combatant will receive a payoff of $[(1 - k)f]/2$, where $k\,(0 \leq k \leq 1)$ is the winner's risk of injury through aggression. Thus, we observe strong "aggressive interactions" between A strategists. The variable k is simply the cost of aggression that can be measured as the probability of a winner's death or injury in an aggressive confrontation between A strategists (an example is shown in Appendix 3).

In the above basic payoff matrix, by introducing the relatedness of interactors, r, we obtained the following "inclusive fitness matrix" (Hines and Maynard Smith 1979):

	C	A
C	$\frac{(1+s)(1+r)f}{2}$	rf
A	f	$\frac{(1-k)(1+r)f}{2}$

From the inclusive fitness matrix, we could identify a possible evolutionary stable strategy (ESS), q^* (the frequency of A strategy, Maynard Smith 1982a), as follows (we can also calculate q^* directly from the basic matrix; Saito and Takada 2009):

The fitness of A strategist males (adopting A strategy) in a population in which all other individuals adopt a mixed strategy at q is denoted as E [A, (mixed strategy, q)] and that of C strategist males as E [C, (mixed strategy, q)].

Then

$$E\left[\mathrm{C},(\text{mixed strategy},q)\right]=\frac{(1+s)(1+r)}{2}f(1-q)+rfq \quad \text{and}$$

$$E\left[\mathrm{A},(\text{mixed strategy},q)\right]=f(1-q)+\frac{(1-k)(1+r)}{2}fq.$$

The candidate ESS frequency, q^*, is obtained when E[A, (mixed strategy, q)]=E[C, (mixed strategy, q)] (Maynard Smith 1982a). We could then get $q^* = [1 - r - s - rs]/[(1 + r)(k - s)]$. After several additional calculations, we confirmed that this formula showed a mixed ESS under $1-s>0$ and $k-s>0$ conditions (Grafen 1979).

Next, let us see how the ESS (frequency of A strategy, $q^* = [(1 - r) - s(1 + r)]/[(1 + r)(k - s)]$) varies with the three parameters, r, k, and s, by simulation. If a risk of aggression, k=0.35 (measured value in Appendix 3) exists, then three ESS types, namely, pure A strategy, mixed strategy, and pure C strategy, can each become ESSs depending upon the r-values at every s-value satisfying $k-s>0$ (Fig. 6.12a). The area of mixed strategy gradually changes with r and s values similar to the skirts of mountain having an "aggression" peak. Thus, we have herein learned that male

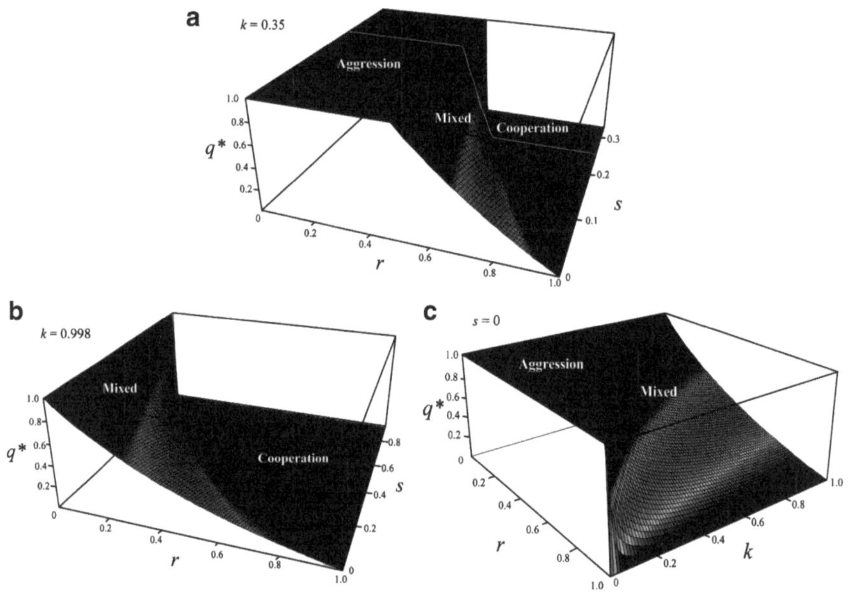

Fig. 6.12 Calculated evolutionary stable strategy (ESS) (frequency q^* of the A strategy) surface with various values of r and s at k=0.35 (**a**) and ESS surface with various values of r and s at extremely high k=0.998 (**b**). Furthermore, q^* gradually changes with k value even when s=0 (**c**)

aggression compromises cooperation at wide ranges of r and s values. Reconfirming the result, under the conditions of $k-s>0$ and $(1-k)/(1+k)<r<(1-s)/(1+s)$, we must observe a mixture of "aggression" and "cooperation" states in a population (note that $r>(1-s)(1+s)$ is the absolute condition under which cooperation will evolve, as shown previously). This result means that we can then observe "clinal variation in aggressiveness with the relatedness (r) of the interacting males" (as expected in Hamilton 1979; Saito 1995a; Saito and Sahara 1999) if the other parameters, k and s, are constant. This is just the thing that we intended to detect by the present model. As such, we learned that "aggression" and "cooperation" are two extreme behaviors of continuous change of the relatedness, r.

Meanwhile, the model provides other several important predictions. In the present model, we can generate several extreme conditions that have been known in previously proposed models:

When $r\approx0$ (although $r=0$ is out of the present model), there are still conditions under which cooperation evolves, if k and s are sufficiently high (Fig. 6.12b). Thus, cooperation would evolve through a parasocial route as emphasized in animals with low kinship (Lin and Michener 1972; Maynard Smith 1982b; Queller 1985; Ito 1993).

When $s=0$, the aggression frequency (q^*) still varies with the value of k (Fig. 6.12c). Male aggression behavior itself shows variation by the effect of k at a constant r, as stated by Maynard Smith (1976) and Enquist and Leimar (1990); this means that variation in male aggressiveness will evolve in connection with the winner's risk of aggression, and sometimes it has no relationship with the change of r (Fig. 6.12c). However, it should be noted that the aggression frequency also varies with r at a constant k, such that there is a condition under which Hamilton's (1979) prediction is satisfied. Furthermore, if aggression behavior results in a catastrophe (at higher k; Fig. 6.12b), then regardless of their relatedness and even if s is low, animals should choose "cooperation" as is known in the hyena, lion, wolf, and so on.

In relationship to the recent theories of cooperation and competition between relatives (Taylor 1992; Queller 1994; West et al. 2002), I have to address how the present case is different from (or similar to) them. Taylor (1992) and Queller (1994) discussed how the benefits of increased relatedness that arise as a result of limited dispersal are exactly canceled out by the cost of increased competition (if they are equally local) between relatives. In fact, West et al. (2001) showed that Hamilton's (1979) rule is not applicable to wingless fig wasp taxa: the level of fighting between males shows no "negative" correlation with the estimated relatedness of interacting males but is negatively correlated with the number of females (mating opportunities) in a fig (Sect. 6.2). As they said, this finding is very consistent with Taylor's (1992) theoretical prediction. Then West et al. (2002) proposed that if we intend to apply Hamilton's rule to some "altruistic" behaviors between relatives, then "effective relatedness" adjusted by the decrease in kin effect that accrues from kin competition should be introduced. For example, if kin individuals cooperate to some extent, but still compete for certain things (e.g., mates, resources, and so on), then the effect of kinship (relatedness) should be discounted. How the kin effect is discounted by such competition depends upon the pattern of dispersal (or viscosity) and the scales at which competition and cooperation occur. Thus, the effective relatedness between individuals should be determined from two kinds of relatedness, i.e., when cooperating and when competing. West et al. (2002) concluded that the reason why there is no relationship between relatedness and male aggressiveness in fig wasp males is that the effect of the relatedness when cooperating is exactly the same as the cost of competing (i.e., the former is completely offset by the latter).

However, we here proposed that there is a condition where aggression compromises cooperation at the same spatial scale through "classic" relatedness. Thus, the effect of individual selection is not always completely offset by kin selection, if we focus on the selection intensities, i.e., by introducing practical parameters, such as k and s. Because there is no rational explanation that the magnitudes of kin and individual selections are equal even at the same spatial level, thus where egoism compromises with cooperation on r must be rather an important question in sociobiology. In other words, there is no need to idealize "effective relatedness" (West et al. 2002) other than classic r, if we can appropriately determine the magnitudes of kin selection and individual selection. The case of the polyembryonic encyrtids reported by Giron et al. (2004) in which soldiers' aggressiveness between lineages is correlated with r may be another example of the present scenario. In these cases, there is no difference in spatial scale between competition and cooperation, but the levels of aggression and cooperation should vary with the relatedness between individuals.

In any case, I believe that the Saito–Takada model is applicable not only to male-to-male interactions but to many kinds of aggression and cooperation in nonrepeated games as well. The conditions required in this model are that there are two different kinds of selection against group-living animals, such as intrasexual selection or resource competition for egoism, and kin selection(s) beneficial to cooperation. These situations frequently occur in animals, such as male-territorial birds and fishes, and the present model may explain why these males do not always fight mortally (Van Hoff 1990). Thus, if we change our focus to multiple selection pressures that operate on both individual lifetime fitness and inclusive fitness, then we may find many examples explainable from the present model in nature.

Fitting Practical Data to the Model

In the previous section, we could show how r determines male behavior in *St. miscanthi* and *St. longus*. Regrettably, we have still no practical data on relatedness from molecular-based methods between interacting males because of the homogeneous nature of *St. miscanthi* genes (we have found several microsatellite loci, but they do not show sufficient variation). Therefore, it is difficult to judge whether the model's prediction is correctly realized in nature. However, if we consider the opposite perspective, we can evaluate relatedness from the Saito–Takada model by substituting the practical parameters that can be obtained experimentally.

The male aggressiveness of *St. miscanthi* was evaluated as the probability of the death (by combat) of one of the two males paired in a woven nest (Sect. 6.1; see Figs. 6.5, 6.6). So as to fit the present model, I translated this as the frequency of A strategy, q, by the procedure shown in Appendix 3. The cost of aggression, k, was estimated from the previous experiments in which two males killed each other simultaneously ($k=0.35$ for HG form; see Appendix 3). The net effect of cooperation, $s=0.25$, was also evaluated as the increase of counterattack success (≈survival rate of offspring; Fig. 3.24) against predators by two males (using the data obtained in LW form; Appendix 3).

Then s and k were substituted in $q^* = (1 - r - s - rs)/[(1 + r)(k - s)]$, and the change of the ESS curve (the frequency of A strategy) with r-values could be simulated (Fig. 6.13; see white line in Fig. 6.12a). Next, setting q^* calculated from the experimental data (Appendix 3) at the vertical axis and the winter minimum temperature (=regarded as relative relatedness between males) at the horizontal axis, I obtained Fig. 6.14. By indexing the plot points of HG (open circles) in Fig. 6.14, I expanded and contracted the horizontal axis of Fig. 6.13 to obtain a curve that fits the almost all population plots on a computer. Finally, I obtained Fig. 6.15, on which the cline of HG form is well fitted. The same procedure was applied for the LW form, providing Fig. 6.16. From these figures, we could estimate the relatedness of males as the values from 0.46 to 0.60 in HG form (Fig. 6.15) and 0.57 to 0.62 in LW form (Fig. 6.16). The aggression frequency cline may form at some point between these values.

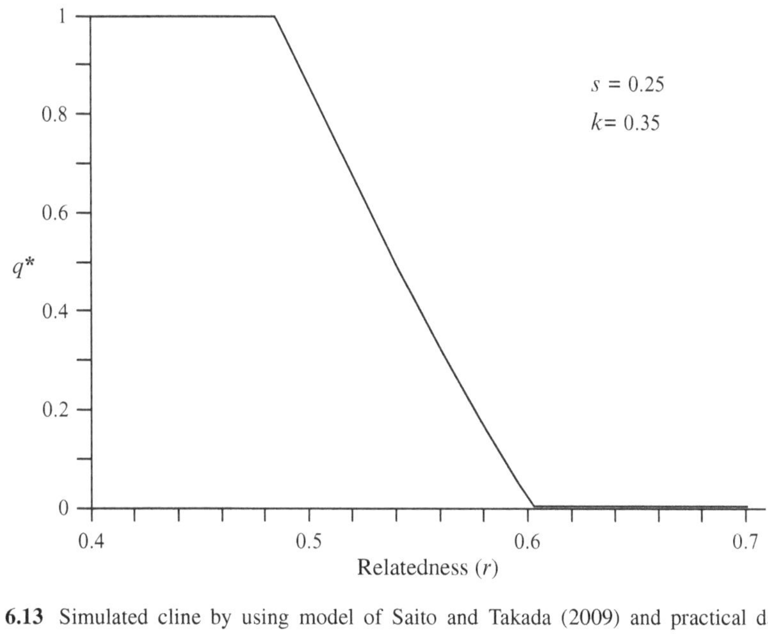

Fig. 6.13 Simulated cline by using model of Saito and Takada (2009) and practical data in Appendix 3

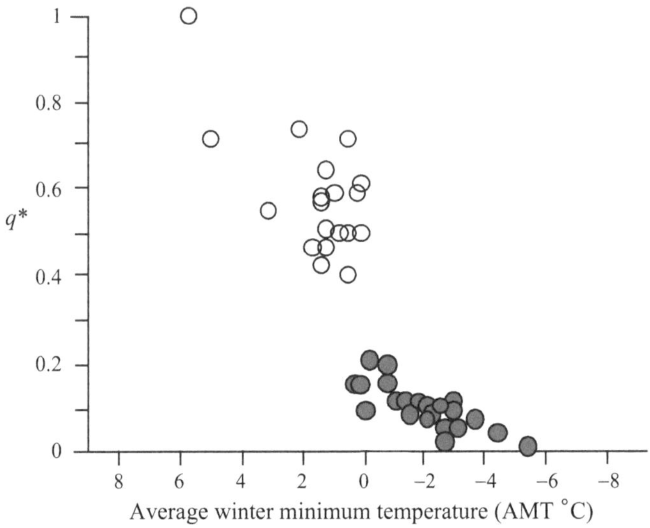

Fig. 6.14 Observed cline in the field (q^* are transformed from Fig. 6.6 data as shown in Appendix 3). Note that the horizontal axis (*AMT*) is reversed in scale (against Fig. 6.6.,6.6) to adjust AMT to the expected change of relatedness, i.e. high temperature at left and low temperature at right

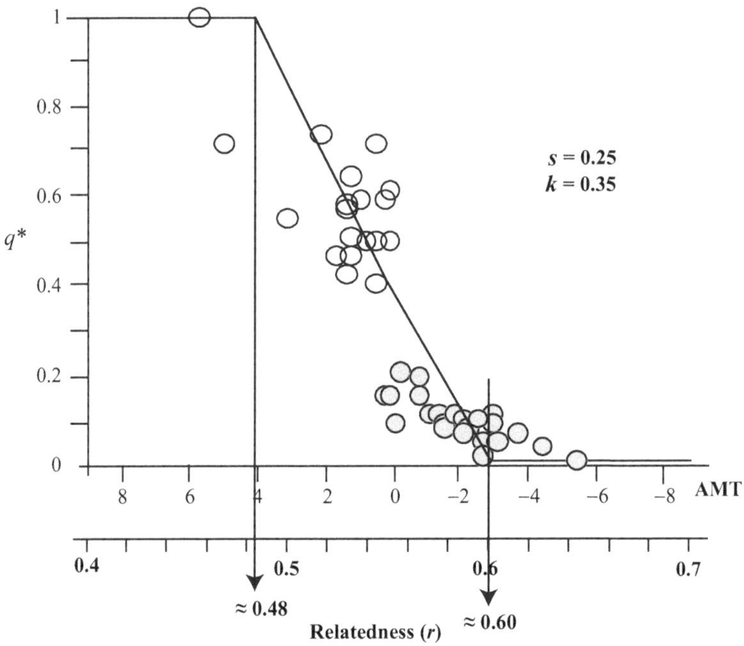

Fig. 6.15 Fitting of EES curve (Fig. 6.13) to the observed cline (*open plots*) in HG form. *AMT* is an index of winter harshness (average daily minimum temperature in winter); r is the relatedness between males, and q^* is the frequency of aggression strategy. *Arrows with numerals* show the estimated r values

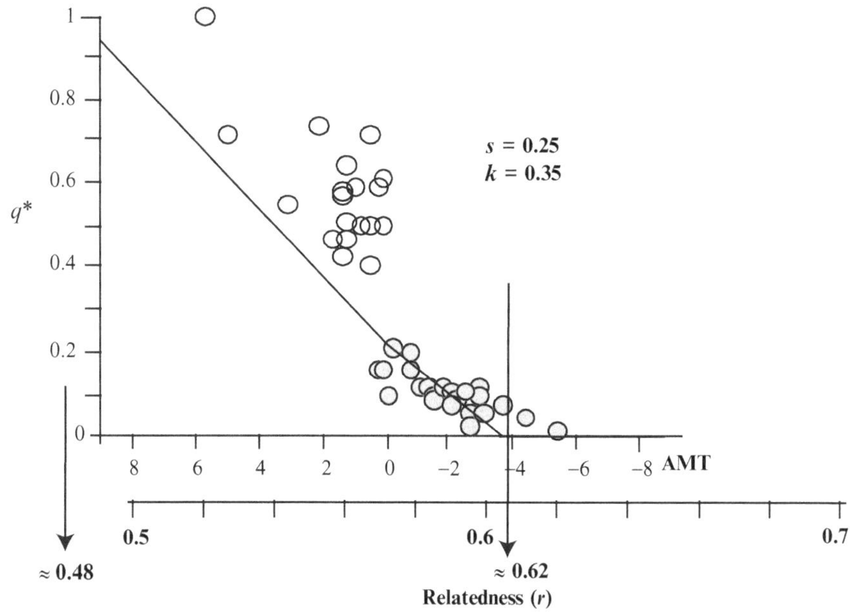

Fig. 6.16 Fitting of EES curve (Fig. 6.13) to the observed cline (*gray plots*) in LW form. AMT is an index of winter harshness (average daily minimum temperature in winter); r is the relatedness between males, and q^* is the frequency of aggression strategy. *Arrows with numerals* show the estimated r values

Thought Exercise

Are the relatedness values ranging from $r=0.48$ to $r=0.62$ at which the mixed strategy (cline) appears in Figs. 6.15 and 6.16 reasonable estimates? As stated before, I have been unable to address this directly, but it is possible to roughly evaluate the r-value by the use of indirect data from field observations. Even though there is a danger of double-quoted hypothesis errors, I am not afraid to do so here. Let us discuss this as a thought exercise only. One available data item is the proportion of unfertilized females in overwintering females in the field. Furthermore, the overwintering probability of males by spring is available. Because *St. miscanthi* is a haplo-diploid organism, there is little relatedness between father and son if they are under outbreeding conditions. On the other hand, if the mites are under inbreeding conditions, the relatedness between males drastically increases (see Box 4.11). Therefore, the structure of the mating population must influence relatedness between males. To estimate the relatedness from such data, we need several presuppositions for simplification, as follows. Assumption 1: There are no overwintering males. Assumption 2: There are no immigrant males (mating occurs only between nest members), such that unfertilized females produce only males and they mate with their sons. Assumption 3: all nests are founded by single females.

I first attempt to evaluate the relatedness between males from the fertilization status of overwintering females only. Under the three assumptions, the average relatedness between males in a nest can be estimated as follows: Between males produced by an unfertilized female in the first generation (produced by overwintered females), $r=0.5$ on average (because they are all sibs), and between males produced by a mother–son mating female in the next (second) generation, $r=0.625$ (Box 4.11). In the other nests where single fertilized females reproduce, the average relatedness between males is 0.5 regardless of the generation. Here let us denote the unfertilized rate as u. The observed proportion of unfertilized females ranged from a minimum of 0.03 in Wakamiya to a maximum of 0.36 in Unzen populations. Then, in the first generation, the relatedness between males in a nest is $r=0.5$ regardless of fertilization status, but in the second generation it will increase to $r= u\times0.625+(1-u)\times0.5=0.503$ in Wakamiya and $r=0.545$ in Unzen populations.

Of course, the above relatedness value must be an overestimation because of assumptions 1 through 3. Relaxing assumption 3, the proportion of nests founded by i foundresses is denoted as v_i ($i=1, 2, 3\ldots$) in spring nests (just after overwintering). The relatedness must decrease depending upon i and v_i, if the relatedness between foundresses is low. Let assume that there is no relatedness (note that this assumption is too strict) between foundresses of a nest ($r=0$, an extreme case); then we can calculate the change of r (under assumptions 1 and 2) between males in the first generation as $r=0.5\sum(v_i/i)$ (Box 6.2). By using v_i observed in several populations (see Fig. 6.4), we can estimate $r=0.29$ in the first generation. In the second generation, the r must increase by the effect of mother–son mating, such that r must increase. However, it is too complex to estimate here. Then, $r=0.29$ must be an underestimation.

Let us try to calculate r from the effect of male overwintering probability (relaxing assumptions 1 and 2 under assumption 3). Overwintering males must affect the relatedness between males in two ways, i.e., they decrease unfertilized females during winter to early spring, and increase outbreeding in spring nests especially in the second generation in spring. Because the former effect primarily changes the rate

Box 6.2 Estimation of relatedness between males

How can we calculate relatedness between males when there are overwintered males in some spring nests? The average relatedness between males produced by i females (all females produce equal number of males) is as follows:

1 female = 0.5, 2 females = 0.5/2, 3 females = 0.5/3 ... 0.5/i.

Because v_i is the proportion of nests founded by i foundresses, thus we can obtain the average relatedness, $r=\sum_i^n \frac{0.5v_i}{i}$

of unfertilized females in spring, thus it is a synonymous case with the first attempt. Here I considered the latter effect only, i.e., no unfertilized females overwinter. How the relatedness between males is affected by overwintering males can be calculated from the proportion of nests where overwintered males exist in the first generation as follows.

If there is no relatedness between overwintered males and the newly produced males in spring nests, and the number of newly born males is assumed to be two on average (males are rare in this species; see Table 4.3), the relatedness of males in a nest where an overwintered male immigrated decreases from 0.5 to 0.167. By denoting the proportion of nests in which an overwintered male exists as m, we can roughly calculate as $r = 0.167m + 0.5(1 - m)$. In the most aggressive population (HG) collected in Ashizuri, 0.81 males per nest were observed in winter (February). On the other hand, only 0.02 males per nest were observed in the Odochi population (LW) that shows lower male aggressiveness. I could assume that 81% of nests involved at most one overwintered male in the former population (remember that only single males can exist in such nests; Sect. 6.1) and 2% of nests in the latter population. Therefore, I obtained $r=0.2$ in Ashizuri and $r=0.493$ in the Odochi population. Because the assumption that there is no relatedness between overwintered males and newly born males must be too severe in this species, these estimations must be underestimations.

As such, the several attempts to estimate r between *St. miscanthi* males interacting in their respective nests may range from 0.2 to 0.55 in spring nests. These estimations are not independent, and one may influence the other. Furthermore, there are expected to be other factors that may increase the mean relatedness between interacting males, such as the primary relatedness (Wright's coefficient of inbreeding) between the foundresses of the first generation and that between overwintered males and newly born males, both of which were assumed here to be zero. Thus it must be noted that the relatedness evaluated here may be a rather "underestimated" value as a "thought exercise." Actually, Antolin (1999) showed that total inbreeding coefficient (F_{it}) in a population of *Trichogramma pretiosum* in a field was 0.246. If we could assume that the value is near to *St. miscanthi* and the relatedness is additive (Lynch and Walsh 1998), the range of relatedness becomes $r=0.45-0.80$. It is noticeable that the range of the relatedness predicted here well overlaps with the range predicted from the Saito and Takada (2009) model, i.e., $r=0.48-0.62$ (see Figs. 6.15, 6.16).

Chapter 7
Is Basic Science the Father of Applied Science?

So far, I have described the behavioral and evolutionary ecology of spider mites, mainly by drawing upon several social spider mite species. As scientists, we sometimes struggle to reconcile social (applied) requirements with our own interests. As I am employed in an agricultural college, I have often struggled with such a dichotomy: How do your studies contribute to agriculture and/or society? Although it is perhaps an old-fashioned example, I intend to describe my limited experiences with this theme.

7.1 What Has Happened in Chinese Bamboo Forests?

In the autumn of 1997, I received a letter from a researcher at the Fujian Plant Protection Institute, China. In that letter, an unbelievable situation was described: Outbreaks of pest mites on moso bamboo (*Phyllostachys pubescens*) had been occurring since the late 1980s, and many bamboo forests in Fujian province had suffered serious damage (Zhang and Zhang 2000). Moso bamboo (giant bamboo) is a species native to China and is thoroughly utilized in industry, housing, and handicrafts, and its shoots are a prized food item in many Asian countries. Furthermore, moso bamboo forests play a vital role in preventing soil erosion in mountainous areas. Spider mite outbreaks are not uncommon in agricultural fields and forests, but the outbreak species are mostly restricted to *Tetranychus*, *Panonychus*, and *Oligonychus* species – all of which are regarded as *r*-selected species (Sect. 2.2) with great potential for population increase. However, the main species of this particular outbreak was *Stigmaeopsis* (*Schizotetranychus*) *nanjingensis*, Dr. Zhang said. This species is a very close relative of *St. longus* and *St. miscanthi*, the life histories and social life style of which I have previously described at length. Several photographs as well as the dried-up leaves covered with dead mites enclosed in the letter suggested this was an extreme situation (Fig. 7.1). The primary cause for my surprise was my belief that *Stigmaeopsis* spp. are *K*-selected species that have evolved in stable habitats with many natural enemies, as described in Sect. 3.6. I am convinced that this is why they evolved sociality.

Y. Saito, *Plant Mites and Sociality: Diversity and Evolution*
DOI 10.1007/978-4-431-99456-5_7, © Springer 2010

Fig. 7.1 Damaged moso bamboo forest in Fujian Province in China (photo by Zhang YX)

Thus I immediately decided to go and observe what was happening in China. In the autumn of 1998, I first visited Fujian Province to observe the pest outbreaks in moso bamboo forests. However, when I visited Fujian Plant Protection Institute, I heard from Dr. Zhang (the researcher who had initially contacted me) that the institute researchers could not reach a consensus about the causes of bamboo forest destruction; i.e., several believed that it was caused by disease, others that it was a result of air pollution. Specialists in plant diseases, entomology, and acarology then conducted a comprehensive survey in several regions of Fujian Province. We subsequently concluded that a "mite outbreak" was the main cause of the moso bamboo forests' destruction.

It was just after my first visit that our joint research initiative between Japan and China aimed at restoring moso bamboo forests began (with initial funding from the Nissei Zaidan and Fujian Government, but with subsequent support from JSPS and MEXT). For almost 10 years we have been conducting surveys in China and Japan, and we have obtained several important results about both the causes of the mite outbreaks and possible ways to restore moso forests.

There are about 275,000 ha moso bamboo forests in Fujian Province, of which approximately 35% were moderately to heavily damaged by phytophagous mites (Zhang et al. 2000). Furthermore, it was suggested that the outbreaks were triggered by a change in bamboo cultivation practice from polyculture to monoculture to increase the yields of bamboo culms and shoots, which inevitably lead to a decrease in both plant and arthropod (pests as well as their natural enemies) diversity (Zhang et al. 2004). Zhang (2002) studied the mite fauna of polyculture moso bamboo forests in comparison with monoculture ones, showing that there is a great difference in the species richness of predatory mites between them. Furthermore, she noticed that the time of the initial mite outbreaks corresponded almost exactly to the time when the moso bamboo plantations in Fujian Province were shifting from polyculture cultivation practices to monoculture ones (Zhang et al. 2004).

Plausible Factors for Pest Outbreaks

Through repeated surveys on the mite fauna in Fujian moso forests, we became more and more inclined to believe that the lack of natural enemies, which would have ordinarily suppressed *St. nanjingensis* populations before the change in cultivation practice, must be a key factor. Figure 7.2 shows the population fluctuations of *St. nanjingensis* and its phytoseiid natural enemies (predators) in heavily damaged moso bamboo forest (monocultures). There are no consistent trends in the dynamics of either prey or predator, showing no effective prey regulation by the predators. If there is a classic-type predator–prey population interaction, the prey population initially increases, then the predator population increases by consuming the abundant prey. The increased predation pressure then causes the prey population to decrease, which in turn leads to a decrease in predator numbers decrease as a result

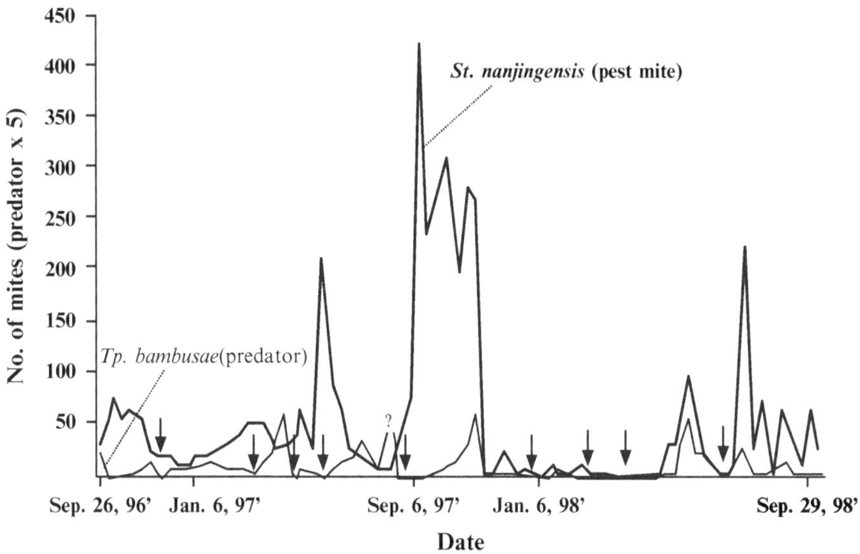

Fig. 7.2 Outbreaks on monoculture Moso bamboo forests in Fujian, China (number of mites per ten leaves). *Arrows* show predator extinction (from Zhang 2002; Saito et al. 2008b)

of a lack of prey. Thus, we may observe a more or less delayed oscillation of the predator population following the prey population dynamics (Lotka 1925; Volterra 1926; Huffaker 1958 in Fig. 2.15). We ascertained that the most abundant predator species occurring on moso bamboo was *Tp. bambusae*. Needless to say, I was surprised, because this predator is known to be a formidable regulator of *Stigmaeopsis* spp. in *Sasa* forests in Japan (Sect. 3.5). Summing up this information, I got the idea that polyculture plantations contain several key plants other than moso bamboo on which other *Stigmaeopsis* species inhabit. These *Stigmaeopsis* species are a vital alternative (secondary) prey for *Tp. bambusae*. An important point is that this predator species only occurs with *Stigmaeopsis* species in Japan (Sect. 3.5), because it has a strong preference for living under *Stigmaeopsis* woven nests and has a bit of difficulty reproducing without such nests (Saito 1990b). Based on these facts, I doubted whether *Tp. bambusae* could persist in monocultured moso bamboo forests.

Approaching a Factor

If so, then what are these key plants? Fortunately we soon got the answer from a survey of an old abandoned moso bamboo forest near the offices of the Fujian Plant Protection Institute. One day when we were taking a rest with cups of Ulong tea in Dr. Zhang's office, her assistant, Mr. Ji, brought a bundle of Chinese silvergrass (*Miscanthus sinensis*) collected in the nearby forest to ask me to identify the spider mites infesting it. At that time, I had no information about whether *Stigmaeopsis miscanthi* was distributed in China. Hence, I was very surprised to see so many injury spots of this mite species on the leaves. This observation later became very important to hypothesize on the origin of the two forms of *St. miscanthi* in Japan (Sect. 6.4), but at that time it was a crucial hint in our search to discover the key causative factor of the *St. nanjingensis* outbreaks in moso bamboo forests. Taking the silvergrass, I observed the leaves one by one under a dissecting microscope, and I found numerous *Tp. bambusae* in the nests of *St. miscanthi*! *Tp. bambusae* is frequently found in *St. miscanthi* nests on Chinese silvergrass in Japan (Chittenden 2002). Furthermore, Chinese silvergrass must be native in China, as expected from its scientific name. Right then, I felt that I could fully understand the cause of mite outbreaks in moso bamboo forests.

After the foregoing discovery, we conducted two experiments to further clarify our hypothesis: (1) Can *St. miscanthi* infest moso bamboo? (2) Can the *Tp. bambusae* collected from *St. miscanthi* really feed on *St. nanjingensis*? We got our answers very quickly: *St. miscanthi* never reproduces on moso bamboo leaves, and *Tp. bambusae* obtained from *M. sinensis* is the same as that obtained from moso bamboo forests and can effectively feed upon *St. nanjingensis* (Zhang unpublished data). Furthermore, Chinese silvergrass was a very common undercover plant in moso bamboo forests, but as it is now used as a material for papermaking, it is periodically harvested and sold.

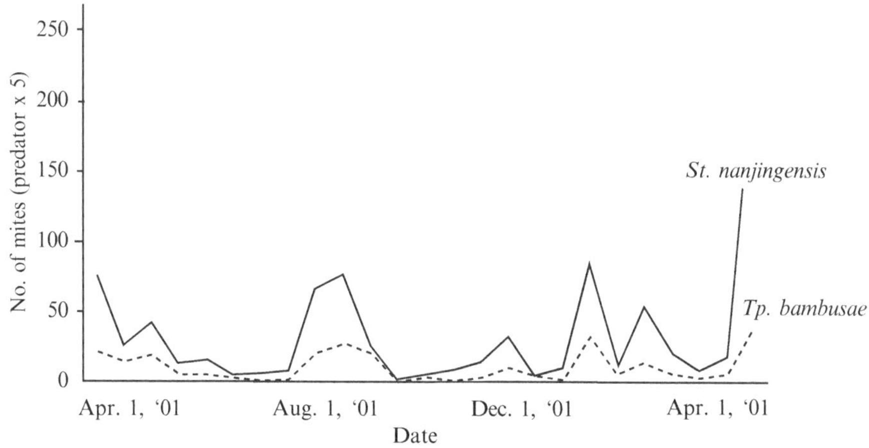

Fig. 7.3 Predator (*Tp. bambusae*) and prey (*St. miscanthi*) populations on Chinese silvergrass (*Miscanthus (M.) sinensis*) growing in a mixed (polyculture moso bamboo) forest (number of mites per 50 leaves) (from Saito et al. 2008b)

As such we had sufficient information about Chinese silvergrass as a nursery plant where *Tp. bambusae* can persist. Then we conducted a 1-year study on the population dynamics of *St. miscanthi* and *Tp. bambusae* (all stages were lumped) on *M. sinensis* (hereafter called MS) in the abandoned moso bamboo forest where *St. miscanthi* was collected. As seen in Fig. 7.3, the *St. miscanthi* population periodically fluctuates and the *Tp. bambusae* population shows similar fluctuations with a short time lag, suggesting that the latter regulated the former population.

7.2 Simulation of One-Predator–Two-Prey Systems

However, whether the existence of nursery plants for *Tp. bambusae* in moso forests really helps regulate *St. nanjingensis* is another problem. From a theoretical perspective, predator–prey interactions have generally been thought to be unstable because the conditions under which the Lotka–Volterra equation is hypothesized are usually hard to find in practice (Kuno 1987). On the other hand, Tansky (1978) and Matsuda et al. (1985) showed that predator–prey systems under which predators can switch from one prey species to another in a prey density-dependent manner easily attain a stable state. In the latter studies, it was hypothesized that predator individuals have the ability to switch between two prey species depending upon respective prey density. Thus, the switching threshold of prey density is regarded as an evolutionary trait of individual predators. We thought that this model situation might be analogous to the systems observed in moso bamboo and Chinese silvergrass in China. Although there is no way for *Tp. bambusae* to switch prey species on a single host plant species, because only a single species occurs on it, they could realize the switch simply by moving to the other nearby host plant to access another prey species. We attempted to test this idea by using a computer simulation (Saito et al. 2008b; Tsuji et al., unpublished data).

The purpose of the simulation described below is to show what happens in pest mite populations when moso bamboo plantations are switched from polyculture to monoculture practices and to define the causal keys to the outbreaks. To ascertain the factors causing the trends observed in moso bamboo (abbreviated as PP) and silvergrass (MS), we developed a simple system model that can simultaneously simulate "one predator–two prey" interactions on two host plant species (Saito et al. 2008b). Most data for *St. nanjingensis* were referred from Zhang and Zhang (2000), those for *St. miscanthi* from Sato et al. (2000), and for the predator, *Tp. bambusae*, from Saito (1990b) and Zhang (2002). In addition, several unknown parameters for prey species were obtained by substituting data for closely related species. Climatic conditions were deemed constant, and all life history parameters were fixed as being under 25±1°C, favorable humidity (50–80% r.h.), and no rain conditions. Furthermore, host plants were all considered to be infinitely abundant, such that there were no food limitations on the two prey species. The ultimate purpose of this simulation model was thus to ascertain both the previously undetermined effect of successful predator migration between PP and MS and the effect of predation success under field conditions on the stability of the overall system.

The program consisted of two distinct host plant subsystems, i.e., MS and PP (Box 7.1). No compatibility in prey species exists between the host plants, with *St. nanjingensis* occurring only on PP, and *St. miscanthi* occurring only on MS, as mentioned before. The predator *Tp. bambusae*, on the other hand, can readily eat both prey species by migrating between the two host plants with particular probabilities of immigration success. We adopted two stages – one is nonreproductive (stage 1; immature females from egg to preovipositional female) and the other is reproductive (stage 2; mature females) – to make the situation as realistic but as simple as possible. The model was defined as difference equations (a more detailed explanation is given by Tsuji et al., manuscript submitted). Strictly defining systems stability in systems simulation is usually difficult, such that we tentatively defined the "stable state" to be when the predator–prey system continued for at least 1,100 days (about 3 years), except for cases when both the predator and prey populations increased indefinitely (overshooting the equilibrium).

First, we simulated whether two subsystems (PP and MS) reach stable states when isolated: initial prey and predator densities are fixed. The parameters of predator immigration success coefficients from one subsystem to the other (from MS to PP and from PP to MS) were 0 (completely isolated), and the predation success coefficients vary artificially (trial-and-error method). Then we got basically three output types with time. If the predation success coefficients were sufficiently high, the prey population became extinct and shortly thereafter the predator population did so as well. If these coefficients decreased, then the predator population became extinct first, and shortly thereafter the prey population entered a phase of overshooting the equilibrium (Box 7.2), or the two populations increased gradually. Figure 7.4a,b shows the result of a simulation where the predator population first

Box 7.1 Brief flow of simulation model in Chinese silvergrass and moso bamboo

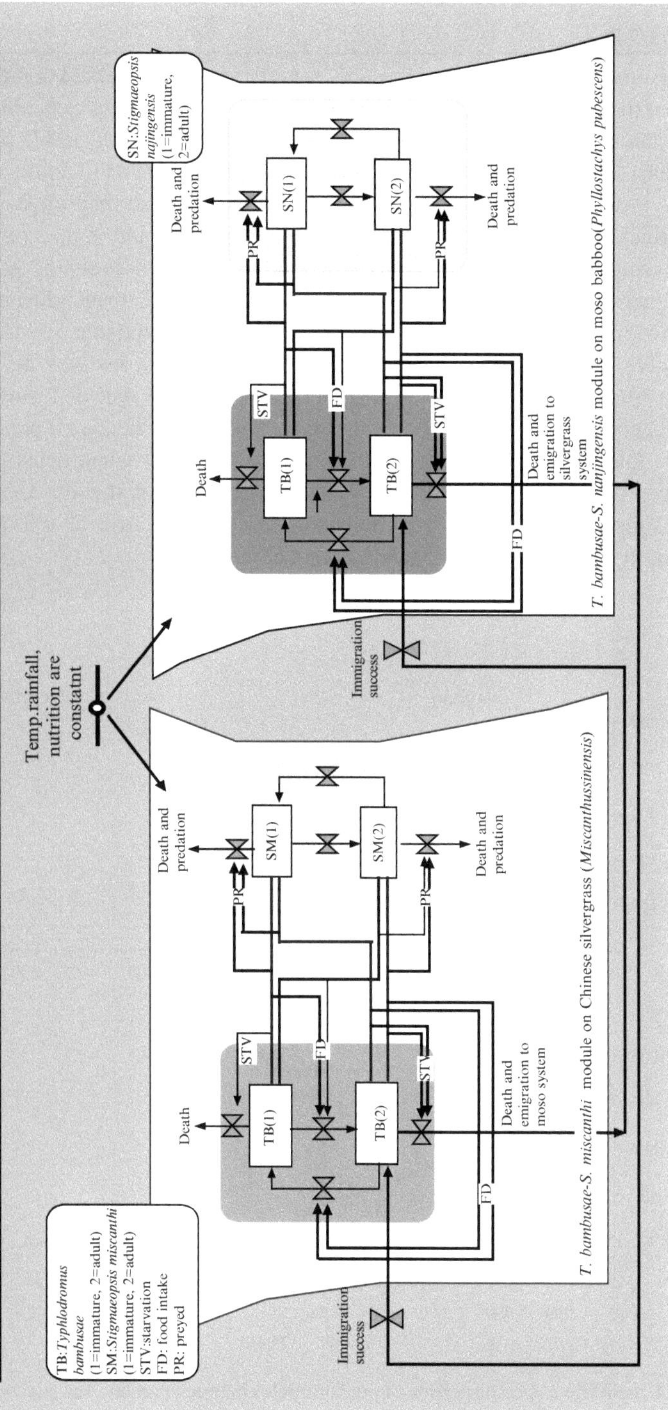

Box 7.2

It is questionable whether the prey population is suppressed completely and then the predator population becomes extinct in nature. In agricultural closed systems, most examples showed such dynamics (Kuno 1987; Saito et al. 1996). However, if such relationships commonly existed in nature, we would not observe any predator–prey relationships, because they all have become extinct. In nature, of course, habitat heterogeneity and some complex relationships must operate to conserve prey and then predator populations. Rather I should point out the famous "life-dinner" principle provided by Dawkins and Krebs (1979). This principle predicts that predators never evolve their hunting ability to overexploit prey at all, because the latter face the risk of losing life but the former have only the risk of losing a dinner. Especially in moso bamboo forests covering several hectares, predators may not able to discover and exploit all prey individuals when densities of both predators and prey become so low. Therefore, we think that it is more realistic to suppose that predators must become extinct because of starvation; then usually some prey individuals can escape predation.

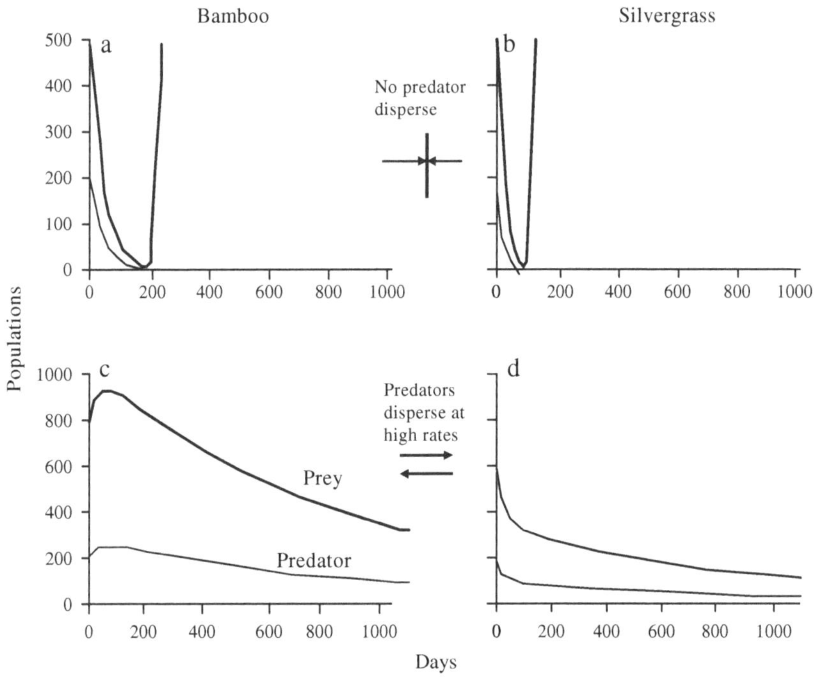

Fig. 7.4 **a**, **b** There is no dispersion of predator mites between bamboo and silvergrass systems. **c**, **d** Predators can move (disperse) at high rates between the two systems (after Saito et al. 2008b)

became extinct and thereafter the prey population entered the overshooting phase. Here we assumed that the predator could not find the prey at low densities because of habitat heterogeneity in a large-scale forest (Box 7.2). These two coefficients were used as constant values in the following simulation.

Next we simulated when the predation success coefficients were fixed at the values in Fig. 7.4a,b and when the migration success coefficients varied arbitrarily. Whether or not both subsystems (=whole system) attained stability was dependent upon the values of the migration success coefficients. There were three outcomes: both subsystems became stable, one system became stable but the other system did not (both predator and prey gradually increase infinitely), and both systems became unstable. Several combinations of the two migration success coefficients can make the systems stable at low levels of prey and predator density (Fig. 7.4c,d).

From these outlined simulation trials, we could infer that a "two-plant–two host-specific pest species–single natural enemy" system can sometimes attain stability through predator migration rates between two plant species. Although this systems simulation is too simplistic and its outcomes too abstract to mimic the actual dynamics of mite populations in Chinese bamboo forests, it at least shows that a "two plants/two pests/single natural enemy" complex (here, *P. pubescens*/*M. sinensis*/*St. nanjingensis*/*St. miscanthi*/*Tp. bambusae*, respectively) can potentially attain stability, suggesting that effective control of pest mite outbreaks might be possible simply by recovering "biodiversity." We now believe that these findings can partly explain the phenomenon that has occurred in Fujian: A change in bamboo cultivation practice from polyculture to monoculture, which simultaneously eliminated Chinese silvergrass vegetation, may have triggered the outbreaks of *St. nanjingensis*.

As such, we have established that reintroducing Chinese silvergrass (*M. sinensis*) is necessary to restore moso bamboo plantations in China. Although traditional polyculture moso plantations involve high plant and arthropod diversity, it is noteworthy that even the recovery of just a single plant species can greatly improve system stability, i.e., sustainability in agricultural fields. Thus, we may show that the conservation of diversified systems must contribute to the stability of biosystems as well as provide biological control of pest species by using endemic natural enemies. In other words, we may prove that biodiversity works as an effective regulator of pests.

So far I have described an attempt to apply the results of basic research to an applied field, i.e., pest management. If I had not had access to any data about *Stigmaeopsis* biology or their predators before the Chinese survey, I might not have detected any way to solve the mite outbreak in moso bamboo forests. Studies on spider mites occurring on *Sasa* still have no practical value from the point of view of agriculture and forestry in Japan (Box 7.3), such that I repeatedly encounter ironic questions from applied researchers, such as "What is the purpose of your study?" and "Is a study on giant panda diet necessary?" I believe that basic science has provided the key ideas for how to solve the pest problems in China.

Box 7.3

The discovery that two prey species on different plants stabilize predator populations may suggest another reason why *Stigmaeopsis* species on *Sasa* never experience outbreaks in Japan. In Sect. 3.5 I described that there are at least two species of *Stigmaeopsis* that have different nest sizes. Such a difference in nest size is possibly related to the switching behavior of *Typhlodromus bambusae* and *Agistemus iburiensis*, as modeled by Tansky (1978). If there are such switching traits in these predators, the predator–prey populations must become stable, as predicted by the model.

7.3 Epilogue: Japanese Sato-yama

I would like to address an epilogue to the Chinese survey in this chapter. Even though we could provide a possible means of regulating pest mite outbreaks in Chinese bamboo plantations by reintroducing biodiversity, the question of why Japanese Sato-yama have been destroyed by the overvegetation of introduced moso bamboo remains unanswered.

Moso bamboo, believed to have been introduced into Japan about 270 years ago from China through the Ryukyu Islands (according to one account, it is 1,200 years ago), is now destroying traditional Japanese forests, called "Sato-yama" (Fig. 7.5). This destruction occurs because poorly maintained bamboo forests rapidly cover over and eliminate other native plant species, which loss subsequently triggers landslides. Cheap imported bamboo shoots from Fujian Province have discouraged Japanese farmers from maintaining their bamboo forests. The biological pest management introduced here, if widely adopted in China, might have further adverse affects on the sustainability of Japanese Sato-yama. Thus, a fundamental conflict exists between Chinese and Japanese bamboo plantations. We have learned that the establishment of sustainable agriculture in one country sometimes causes unexpected problems for sustainability in another.

To resolve this problem, we must identify the mechanisms that prevent moso bamboo from covering over and eliminating other plant species in its native areas, such as Fujian Province in China. During the moso bamboo study, I learned that there are two bamboo rat species that feed on moso bamboo shoots. Furthermore, in the old days giant pandas were prolific consumers of bamboo shoots, even in Fujian Province (Fig. 7.6). How these species regulated moso bamboo in the past is now difficult to ascertain, but it is very suggestive that bamboo rats prefer to use Chinese silvergrass for their den sites (Fig. 7.6; Xu 1989). As the introduction of exotic pest species (mites, bamboo rats) carries inherent danger to indigenous flora and fauna, we cannot recommend such a course of action in Japan, however. Thus, conservation of Japanese Sato-yama through the prevention of unchecked moso

Fig. 7.5 Moso bamboo climbing up traditional Japanese "Sato-yama" (Karatsu City, Kyushu district, Japan)

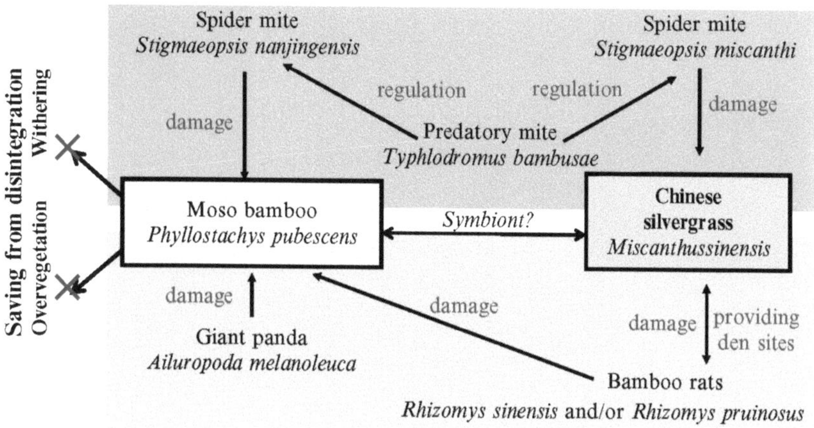

Fig. 7.6 Possible relationships between moso bamboo and Chinese silvergrass and their herbivores and predators in ancient China (from Saito et al. 2008b)

bamboo growth is a serious and difficult problem. Instead, we strongly recommend the heavy exploitation of moso bamboo both as materials for industry and food and as a source of bio-energy (Saito et al. 2008b).

Chapter 8
Further Research for Mite Behavioral and Evolutionary Ecology

Evolutionary ecology and sociobiology are sciences for studying the selection pressures operating in nature. As I have described, I believe that spider mites and phytoseiid mites are ideal materials for such studies, because they are often directly exposed to natural selection. Their short life spans and our ability to observe their entire life cycle in small experimental areas are also great advantages. By these and other merits, we have been able to conduct much of the research described so far.

However, there are also some difficulties involved in using mites for ecological studies. Each trait of the species that we focus on, the nesting behavior of *St. longus*, for example, has to be checked against every kind of selection, i.e., natural selection operating on survival and reproduction, sexual selection on mating success, and synergistic and/or kin selection on social living, all of which must be operating simultaneously in natural habitats. The minute bodies of mites often make it impossible to observe individual lifetime success in the field. Furthermore, environmental conditions surrounding such minute organisms may be quite different from our own sense of realism, as pointed out by Schmidt-Nielsen (1984) and Saito and Suzuki (1987). Difficulties underlying the study of spider mite genetics must be also mentioned. Saito et al. (2000b) revealed that there are recessive genes that govern the depression of female fecundity (Chap. 5). However, they could not determine how many genes (alleles and loci) are responsible for this trait or how intense this effect is in comparison with the wild type. To acquire such information, we have to observe the fitness changes between generations under constant conditions without environmental variance. Because spider mites are phytophagous, it has been (and still is) difficult to keep host plant conditions constant. Therefore, we still have only limited means of obtaining this important information. It is necessary to develop some kind of artificial diet that can keep the food conditions of spider mites constant to overcome such a hurdle.

As is well known in many insects, as well as some mites, chemical cues are a very important means for recognizing the external world (e.g., Dicke et al. 1998; Sato and Saito 2006). This importance partly comes from the fact that mites basically live in a world they cannot see. Are there any possibilities that they might use sound as a cue for communication? Because of their tiny bodies, such a study has never been attempted. Although it may be difficult to separate

Y. Saito, *Plant Mites and Sociality: Diversity and Evolution*
DOI 10.1007/978-4-431-99456-5_8, © Springer 2010

sound from tactile stimulus, we must focus more attention on such possibilities as well as on chemical cues.

I have described several unresolved questions in Chaps. 5 and 6. Among these, the origin of the two forms of *St. miscanthi* is our current research focus. In relation to this theme, Sato et al. (2008) recently reported that there is a contact zone of these two forms and a possibility of interbreeding between them. If such interbreeding frequently occurs in the field, "introgenesis" between the two forms may be expected. Such "hybrid gene effects" may simply explain the variation in male aggressiveness of *St. miscanthi* (M. Kawata and Y. Sato, personal communication) and represent the antithesis of my kin selection hypothesis. However, I believe that this cannot be an alternative to kin selection, because the former is a proximate and the latter is an ultimate factor to explain such a phenomenon. A more important question is why two forms showing "different" intensities of male aggressiveness exist in Japan, even though we know there is a quantitative difference in diapause attributes between the forms (Sect. 6.4). To solve this problem, we have to both widen our perspective, to the Asian continent, for example, while at the same time narrowing it down to embrace the molecular level.

Finally, I address our recent attempts to understand the structure and function of mite communities. *Sasa* bamboo is known to harbor a complex mite community (Chittenden 2002; Mori and Saito 2004). We have only focused on several spider mite species on *Sasa* bamboo, but at least nine species of spider mites and perhaps an equal number of predacious mite species live either syntopically or allotopically in the same forests. I do not know whether the mite diversity observed on *Sasa* bamboo is extraordinary or not, but the reasons why there are so many host-specific species should be explained from the perspective of evolutionary community ecology. The importance of the diversity of organisms has been stressed throughout this book, but do we really know the reasons why there are such complex communities? The relationships between spider mites and their natural enemies are very easy to observe under seminatural experimental conditions as described in this book, such that studies on the complex community existing on *Sasa* from the aspect of evolutionary and behavioral biology may provide an answer to this very important question.

Afterword

After I had completed the draft of this book, I sincerely thought that I had only written about our own studies on spider mites and their natural enemies. This is a concern of the advisers for this book. Anyway, how do you feel as a reader of this book? Some have criticized that such narrow and fanatical research is not scientific, and others have said that scientific research should be made with working hypotheses to be proved before the start. On the other hand, mite researchers may say that I have to write in a more detailed and comprehensive manner. These comments all concern my style of ecological research methodology. Current ecology and evolutional biology have strong trends that hypotheses or theories should come first and then research be conducted to solve or prove them. This is how I have taught my graduate students. However, I do not know when I first developed such habits, but I have tended to adopt rather traditional natural history methodology, which is now considered to be old-fashioned. This is the reason why I define my studies as Fabre like, even though my contributions are hardly comparable to his great works. However, it is apparent that no biological study can be conducted without observing "phenomena" in nature. Therefore, I think that theoretical and or hypothesis verification studies can only be conducted after we observe phenomena that need to be resolved. In this context, ecology and evolutionary biology are inductive. In this book I have introduced several fascinating researchers who have detected general results by using mites, which can then be applied to "zoology." However, if they also pay more attention to mites as "actual animals" to be observed in relation to their evolutionary histories, we are bound to learn more deep insights from their results, I think. From the hypothesis verification style, my contributions are not so general and must be restricted to a narrow area of zoology as well as acarology. I now believe that the dichotomy of scientific methodology mentioned above is unimportant. Theoretical studies should always be accompanied by natural history studies and vice versa. Therefore, if we develop ecology, behavioral, and evolutionary biology, we will always be paying attention to the traditional zoological sciences – taxonomy, phylogeny, and physiology – as well as to theoretical biology. If what I have presented in this book successfully resembles something like this type of integrative biology, I will have achieved my goals.

Appendices

Appendix 1. Brief History of Reinstatement of *Stigmaeopsis*

One of the major difficulties involved in reaching the reinstatement of the genus *Stigmaeopsis* arose from the condition of the type specimen made by Banks (1917). I asked the Smithsonian Institution (USA) to render type specimens of *Stigmaeopsis* (*Schizotetranychus*) *celarius* and had the opportunity to observe them (by courtesy of Dr. Ochoa R). Regrettably, the holotype and paratypes had disintegrated beyond recognition, making morphological observations impossible. This is an unfortunate and inevitable problem in mite taxonomy, because until Saito *et al.* (1993) developed a new method for making permanent specimens using Canada balsam, no way to prepare permanent specimens had existed (Jeppson *et al.* 1975). Furthermore, the drawings made by Banks (1917) for species description were not always accurate and made it difficult to identify characters. On the other hand, the descriptions by Pritchard and Baker (1955) were very detailed and we could compare our specimens with their drawings. We then decided to assume that *Stigmaeopsis celarius* described by Banks was the same as Pritchard and Baker's (1955) description of *Schizotetranychus celarius* (herafter *Sc*). Under this assumption, we could compare all populations of *Sc* distributed in Japan (Fig. 3.11) and China as follows: the LM form is a synonym of *Sc* (redescribed by Pritchard and Baker 1955), L and MS are different species from *Sc* as reported in Saito (1990a), and S and M are also different from *Sc*, L, and MS in their morphologies. However, the question remained as to whether the five Japanese species and two Chinese species should be categorized into the genus *Schizotetranychus* or into the deleted genus *Stigmaeopsis*, because we had also discovered an important difference in the chaetotaxy of the distal segment of the palpus: In all seven species, the distal segment has two simple setae in both females and males, but it has three simple setae in the other species of *Schizotetranychus* as well as in most genera of Tetranychini (Lindquist 1985). Thus, we concluded that such a difference must be an important criterion for separating these seven species from the genus *Schizotetranychus*. Although there is no information about this difference in Banks's (1917) first description, we chose to reinstate the genus *Stigmaeopsis* and determined that these seven species should belong to it. As such, we described M and S as two new species, *Stigmaeopsis takahashii* and *Stigmaeopsis saharai*, respectively, and moved the other five species, including the two Chinese species (*Schizotetranychus nanjingensis* and *Schizotetranychus tenuinidus*), to the genus *Stigmaeopsis*.

After reinstating *Stigmaeopsis*, we faced another problem. Takahashi (1987) observed a species from Thailand that had a similar web nest habit (WN life type) to *Stigmaeopsis* spp. and reported that the chaetotaxy of the distal segment of the palpus was different from the Japanese *Stigmaeopsis* (*Sc* complex) species. We subsequently rediscovered such a difference between *Stigmaeopsis* and the other Tetranychini, as mentioned before. Recently, we (Sakagami et al. 2009) conducted a survey of mite fauna in Thailand and could collect several species with WN life types (*Schizotetranychus saitoi*, *S. approximatus*, and *S. laevidorsatus*). Using these specimens and *Stigmaeopsis* spp., we conducted a molecular phylogenic study and revealed that *Stigmaeopsis* spp. and the Thailand species inhabiting *Bambusa* spp. and *Gigantochloa* spp., form two distinct but very close clades. Furthermore, we found that dorsal chaetotaxy of the Thailand species (parallel-arranged dorsocentral setae) more closely resembles the drawings of Banks (1917) than *Stigmaeopsis* spp. (all species of this genus have dorsocentral setae arranged in a V-shape). This was puzzling. We had no way to continue our taxonomic studies because of the lack of an observable holotype. However, we now think that the *Stigmaeopsis celarius* described by Banks (1917) is one of these three species from Thailand. There is no way of deriving a correct answer from this confusion, but I am now considering whether we must change the present *Stigmaeopsis* into a new genus, then return *Stigmaeopsis* to the Thailand species some time in the future.

The point that I wish to convey from this story is the importance of a holotype. The main difficulty of our taxonomic studies on *Stigmaeopsis* was caused by the disintegration of specimens because of the shortcomings of Hoyer's solution (a kind of gum chloral fluid), the only mounting medium that was available at the time for mite slide preparation. This water-soluble medium is very useful for short-term preparations but has a short life span (10 years at most using sealing plastic). To preserve such specimens periodic remounting is essential, but because of the decreased numbers of taxonomists and curators in universities and museums around the world, this is not always possible. I strongly disagree with this trend, believing instead that a viable means of making permanent specimens is always required before beginning taxonomic study. Now that we have the new Canada balsam methods (Saito et al. 1993) to prepare all type specimens as permanent specimens, spider mite taxonomy faces no such constraints.

Appendix 2. Model of Repeated Blind Sampling

(by Y. Saito, K. Mori, K. Ito, and A. Yamauchi)

In order to adapt the repeated blind sampling model for the nest-scattering trait of *Stigmaeopsis* species in Saito et al. (2008a), we formulated a mathematical model.

Model

The many small *Stigmaeopsis* nests randomly scattered over the leaf surface are considered to be a situation analogous to the tricking effect of prey dummy nests, because under natural conditions such nests persist on leaves for a certain period even after suffering predation. We developed a simple model to show how void nests can effectively trick predators. From the VTR (video tape recorder) and field observations, the conditions assumed for the tricking model are as follows:

1. A prey female makes x nests randomly over a certain area throughout her lifetime.
2. A prey female reproduces N eggs throughout her lifetime.
3. A prey female evenly oviposits N/x eggs in each nest ($N \geq x$).
4. A predator female finds all prey nests with equal probability regardless of the number of nests or the presence of prey eggs within them.
5. When a predator female finds a nest, it always intrudes into the nest regardless of the presence of prey eggs and eats all available prey eggs.

Assuming that i nests have already been found and the prey eggs within them have been consumed [i.e., there are i already preyed (void) nests], the number of nests that have not been found is $x-i$. Thus, the expected number of searching attempts required to find a "nest with prey eggs" is $x/(x-i)$, but $x \neq i$.

From the above, the number of trials (R) in which a predator has found k valid nests ($k \leq x$) is

$$R = x\sum_{i=0}^{k-1}\left[1/(x-i)\right] \tag{A.1}$$

The searching time required to find a nest (regardless of whether it is valid or void) was denoted by s (including the handling time required to intrude a nest). The time required for a predator to eat an egg is denoted by e. Introducing these parameters into (A.1), we obtain the time (T) required for a predator to eat all eggs (N) as the following equation (the resting time including oviposition time must occupy a certain proportion of the total time spent but is ignored here):

$$T = sx\sum_{i=0}^{x-1}\left[1/(x-i)\right] + eN \tag{A.2}$$

From the point of view of the prey, x is a strategy by which prey fitness will increase with time T, because as T lengthens, the fewer prey eggs will be eaten.

Table A.1 Behavioral parameters of *Agistemus (Ag.) iburiensis* at x (number of nests) = 6 determined by VTR observations (n = 5) for 24 h

Parameters	Values estimated (h)
Searching and handling time required for a nest, s	1.27
Time for eating an egg, e	0.33

Appendix 3. Calculation of Practical Parameters in Game Model

By using the raw data without any adjustment (therefore, the data are not consistent with Figs. 6.5 and 6.6), q was calculated as follows: One male death observed experimentally during 5 days (most results were settled within 3 days) is regarded to occur on the following three occasions: the combat between A strategy and A strategy, that between A strategy and C strategy, and that between C strategy vs. A strategy. The frequency of A strategy could be calculated. Let us denote the frequency of A strategy as q and that of C strategy as p. The frequency of A vs. A confrontations is q^2, that of C vs. C is p^2, that of A vs. C is pq, and that of C vs. A is pq. Considering that all the confrontations including A strategy result in the death of one of the paired males, the frequency of p^2 can be calculated from the experimental data as $p^2 = 1 -$ "the probability of the death of one of the two males in the previous studies." Thus the frequency of A strategy can be obtained as $q = 1 - p$.

The aggression cost between A strategists can also be calculated by using the above calculation as follows: The probability of killing each other was 0.081, observed in the Sumoto population of *S. miscanthi*, showing 0.76 probability of being killed by the other male. However, this is not the required value of k, because k is only the risk when males adopting A strategy fight each other. The frequency of A vs. A fighting can be calculated from the frequency of C vs. C (p^2) as above. For example, in the Sumoto population, the frequency of $p^2 = 1 - 0.756$; thus I obtained $p = 0.494$. Because $p + q = 1$, then $q = 0.51$ and $k = 0.081/q^2 = 0.31$. I have also calculated k for several populations in the same manner and obtained values from 0.1 to 0.35 (Saito, unpublished data). Because these estimations are derived from the extreme results of male fighting, they do not include any injury risks that disable the winner, such that it is reasonable to adopt the maximum value of $k = 0.35$.

The net effect of cooperation was newly estimated as "defense success against a young predator by two males"/"that by a single male" in a nest with one female. *St. miscanthi* and its predator species, *Neoseiulus (N.) womersleyi* (this species is one of the predators frequently observed in *St. miscanthi* LW form nests), were collected from Mt. Unzen, Nagasaki Prefecture, Japan, on August 12, 2004 (Saito, unpublished data). Two young males and two unfertilized females of *St. miscanthi* were introduced onto a detached host leaf (*Miscanthus sinensis*), and mating occurred. Four days after their introduction, one female, one male or two males, and eggs (4-12) were left in a nest, and the excess males and females were removed. A predator larva was then introduced into the *St. miscanthi* nest. Three days after the predator introduction, success of offspring defense by *St. miscanthi* adults was observed ($n = 23$~27). The reason why we used *St. miscanthi* LW form for this experiment came from the difficulty of keeping two HG males in a nest simultaneously because of their strong male-to-male aggressiveness. As stated before, "aggression between males" and "counterattack ability against predators" are independent (remember the *St. longus* case), such that we assumed that the data obtained for the LW form could be commonly used for the HG form.

When the predator was killed, the defense was considered to be complete (criterion 1). Furthermore, if the predator was excluded from the nest at every observation (3 times) and if half of the prey immatures survived for 5 days, the counterattack was considered to be successful (criterion 2). By the use of these criteria, the defense success of a single male is 0.130 (criterion 1) and 0.522 (criterion 2), and that of two males is 0.148 (criterion 1) and 0.609 (criterion 2). The net effect of cooperation, s, is calculated as 0.14 at criterion 1 and 0.17 at criterion 2. If we considered the long-term effect [e.g., assuming double predatory intrusion events, $t=2$ (for 10 days); Box 3.4], s is calculated as 0.29 at criterion 1 and 0.36 at criterion 2. Here I used $s=0.25$ as the middle value of 0.14 to 0.36. These estimations are considered to be tentative because of the small number of replications, i.e., $n=23$ (one-male defense) to 27 (two-male defense).

References

Abe J, Kamimura Y, Ito H, Matsuda, H, Shimada M (2003b) Local mate competition with lethal male combat: effects of competitive asymmetry and information availability on a sex ratio game. J Evol Biol 16:607–613

Abe J, Kamimura Y, Kondo N, Shimada M (2003a) Extremely female-biased sex ratio and lethal male–male combat in a parasitoid wasp, *Melittobia australica* (Eulophidae). Behav Ecol 14:34–39

Alcock J (1979) Animal behaviour: an evolutionary approach, 2nd edn. Sinauer, Sunderland, MA

Antolin MF (1999) A genetic perspective on mating systems and sex ratios of parasitoid wasps. Res Popul Ecol 41:29–37

Aponte O, McMurtry JA (1997) Damage on 'Hass' avocado leaves, webbing and nest behaviour of *Oligonychus perseae* (Acari: Tetranychidae). Exp Appl Acarol 21:265–272

Atmer W (1991) On the role of males. Anim Behav 41:195–205

Aviles L (1997) Causes and consequences of cooperation and permanent sociality in spiders. In: Choe JC, Crespi BJ (eds) The evolution of social behaviour in insects and arachnids. Cambridge University Press, London, pp 476–498

Banks N (1917) New mites, mostly economic (Arach., Acar.). Entomol News 28:193–199

Begon M, Harper JL, Townsend CR (1999) Ecology: Individuals, populations and communities, 3rd edn. Blackwell Publishing, Oxford

Bernstein C (1984) Prey and predator emigration responses in the acarine system *Tetranychius urticae – Phytoseiulus persimilis*. Oecologia (Berl) 61:134–142

Blazak C, Ehrensberger, Schuster R (1990) Beiträge zur Kenntnis der Lebenweise der Litoralmilbe *Macrocheles superbus* Hull, 1918 (Acarina: Gamasina). Osnabrücker naturewiss Mitt 16:51–62

Boczek J (1959) Biologia I ekologia sierposza rozkuszkowka (*Cheyletus eruditus*) (Schrank, 1781) (Acarina, Cheyletidae). Press Naukowe IVR 175–230

Bolland HR, Gutierrez J, Flechtmann CHW (1998) World catalogue of the spider mite family (Acari: Tetranychidae). Koninklijke Brill NV, Leiden

Boudreaux HB, Dosse G (1963) The usefulness of new taxonomic characters in females of the genus *Tetranychus* Dufour (Acari: Tetranychidae). Acarologia 5:13–33

Bull JJ (1983) Evolution of sex determining mechanisms. Benjamin/Cummings, Advanced Book Program, Palo Alto, California

Calvert WH, Hedrick LE, Brower LP (1979) Mortality of the monarch butterfly (*Danaus plexippus* L.): avian predation at five overwintering sites in Mexico. Science 204:847–861

Carson R (1962) Silent spring. Houghton Mifflin, Boston

Chant DA (1959) Phytoseiid mites (Acarina: Phytoseiidae). Part I. Bionomics of seven species in southeastern England. Part II. A taxonomic review of the family Phytoseiidae, with descriptions of thirty-eight new species. Can Entomol 91 (suppl) 12:1–166

Charlesworth D, Charlesworth B (1987) Inbreeding depression and its evolutionary consequences. Annu Rev Ecol Syst 18:237–268

Chittenden AR (2002) Diversity in predator–prey interaction of plant-inhabiting mites (Acari: Tetranychidae, Phytoseiidae and Stigmaeidae). PhD dissertation, Hokkaido University, Sapporo

Chittenden AR, Saito Y (2001) Why are there feeding and nonfeeding larvae in phytoseiid mites (Acari, Phytoseiidae)? J Ethol 19:55–62

Chittenden AR, Saito Y (2006) Tactile crypsis against non-visual predators in the spider mite, *Aponychus corpuzae* Rimando (Acari: Tetranychidae). J Insect Behav 19:419–428

Cloutier C, Johnson S (1993) Interaction between life stages in a phytoseiid predator: western flower thrips prey killed by adults as food for protonymphs of *Amblyseius cucumeris*. Exp Appl Acarol 17:441–449

Cone WW (1979) Pheromones of Tetranychidae. In: Rodriguez JG (ed) Recent advances in acarology, vol 2. Academic Press, New York, pp 309–317

Crespi BJ (1994) Three conditions for the evolution of eusociality: are they sufficient? Insect Soc 41:395–400

Crozier RH (1979) Genetics of sociality. In: Hermann HR (ed) Social insects, vol 1. Academic Press, New York, pp 223–286

Crozier RH (1985) Adaptive consequences of male-haploidy. In: Helle W, Sabelis MW (eds) Spider mites. Their biology, natural enemies and control, vol 1A. Elsevier, Amsterdam, pp 201–222

Dawkins R, Krebs JR (1979) Arms races within and between species. Proc R Soc Lond B 205:480–512

Dicke M, Sabelis MW. (1988) How plants obtain predatory mites as bodyguards. Neth J Zool 38:148–165

Dicke M, Sabelis MW, Takabayashi J, Bruin J, Posthumus MA (1990) Plant strategies of manipulating predator–prey interactions through allelochemicals: prospecters for application in pest control. J Chem Ecol 16:3091–3118

Dicke M, Takabayashi J, Posthumus MA, Schutte C, Krips OE (1998) Plant-phytoseiid interactions mediated by herbivore-induced plant volatiles: variation in production of cues and in responses of predatory mites. Exp Appl Acarol 22:311–333

Donzé G, Guerin PM (1994) Behavioural attributes and parental care of *Varroa* mites parasitizing honeybee brood. Behav Ecol Sociobiol 34:305–319

Dosse G (1961) Uber die Bedeutung der pollennahrung fur *Typhlodronus* (*T.*) *pyri* Scheuten (= *tiliae* Oud.) (Acari, Phytoseiidae). Entomol Exp Appl 4:191–195

Dosse G (1967) Beitrage zum Diapause-Problem von *Tetranychus urticae* Koch und *Tetranychus cinnabarinus* Boisduval-Komplex im Libanon (Acarina, Tetranychidae). Pflanzenschutz Ber 34:129–138

Dugatkin LA (1997) Cooperation among animals. An evolutionary perspective. Oxford University Press, New York

Ehara S (1957) On three spider mites of *Schizotetranychus* from Japan. J Fac Sci Hokkaido Univ Ser 6 Zool 13:15–23

Ehara S (1993) Colored guide to the plant mites of Japan. (Ehara S ed), Zunkoku Noson Kyouiku Kyokai, Tokyo (in Japanese)

Ehara S (1999) Revision of the Spider Mite Family Tetranychidae of Japan (Acari, Prostigmata). Species Divers 4(1):63–141

Ehara S, Gotoh T (1992) Descriptions of two *Panonychus* spider mites from Japan with a key to species of the genus in the world (Acari: Tetranychidae). Appl Entomol Zool 27:107–115

Ehara S, Okada T, Kato H (1994) Contribution to knowledge of mite family Phytoseiidae in Japan (Acari: Gamasina). J Fac Educ Tottori Univ 42:119–160

Ehara S, Shinkaji N (1975) An introduction to agricultural acarology. Zenkoku Noson Kyoiku Kyokai, Tokyo (in Japanese)

Ehara S, Shinkaji N (1996) Principles of plant acarology. Zenkoku Nouson Kyoiku Kyokai, Tokyo (in Japanese)

Enquist M, Leimar O (1990) The evolution of fatal fighting. Anim Behav 39:1–9

Faraji F, Janssen A, van Rijn PCJ, Sabelis MW (2000) Kin recognition by the predatory mite *Iphiseius degenerans*: discrimination among own, conspecific and heterospecific eggs. Ecol Entomol 25:147–155

Fisher RA (1930) The genetical theory of natural selection. Oxford University Press, Oxford

Fleschner CA, Badgley ME, Ricker DW, Hall JC (1956) Air drift of spider mites. J Econ Entomol 49:624–627
Foster WA, Treherne JE (1981) Evidence for the dilution effect in the selfish herd from fish predation on a marine insect. Nature (Lond) 293:466–467
Fournier V, Rosenheim JA, Brodeur J, Laney LO, Johnson MW (2003) Herbivorous mites as ecological engineers: indirect effects on arthropods inhabiting papaya foliage. Oecologia (Berl) 135:442–450
Fowler HG, Gobbi N (1988) Cooperative prey capture by an orbweb spider. Naturwissenschaften 75:208–209
Frank SA (1985) Hierarchical selection theory and sex ratios. II. On applying the theory, and a test with fig wasps. Evolution 39:949–964
Fransz HG (1974) The functional response to prey density in acarine system. Center for Agriculture Publication Documents, Wageningen
Furuhashi K, Nishino M (1978) Studies on the dispersal of citrus red mite, *Panonychus citri* McG. Bull Shizuoka Pref Citrus Exp Stn 14:29–42 (in Japanese with English summary)
Fujimoto H, Takafuji A (1993) Oviposition of diapause-induced females of *Panonychus mori* Yokoyama (Acari: Tetranychidae) Appl Entmol Zool 28:106–108
Gerson U (1979) Silk production in *Tetranychus* (Acari: Tetranychidae). In: Rodriauez JG (ed) Recent advances in acarology, vol 1. Academic Press, New York, pp 177–188
Gerson U (1985) Webbing. In: Helle W, Sabelis MW (eds) Spider mites. Their biology, natural enemies and control, vol 1A. Elsevier, Amsterdam, pp 223–232
Gerson U, Aronowitz A (1981) Spider mite webbing. Part IV: The effect of acaricides on spinning by the carmine spider mite *Tetranychus cinnabarinus* (Boisduval). Pestic Sci 12:211–214
Giron D, Dunn DW, Nardy ICW, Strand MR (2004) Aggression by polyembryonic wasp soldiers correlates with kinship but not resource competition. Nature (Lond) 430:676–679
Glennis EJ, Fewell JH, Gadau J, Johnson RA, Larrabee D (2002) Genetic determination of the queen caste in an ant hybrid zone. Proc Natl Acad Sci U S A 99:8157–8160
Goka K (2001) Geographical variation. In: Aoki J (ed) Biology of ticks and mites. University of Tokyo Press, Tokyo (in Japanese)
Goka K, Takafuji A (1990) Genetical studies on the diapause of the two-spotted spider mite, *Tetranychus urticae* Koch (1). Appl Entomol Zool 25:119–125
Goka K, Takafuji A (1991) Genetical studies on the diapause of the two-spotted spider mite, *Tetranychus urticae* Koch (2). Appl Entomol Zool 256:77–84
Gotoh T (1984) Ecological studies on diapause in Tetranychidae, with special reference to diapause attributes in relation to host plant phenology. PhD dissertation, Hokkaido University (in Japanese)
Gotoh T (1986) Local variation in *Panonychus akitanus* Ehara (Acarina: Tetranychidae). Exp Appl Acarol 2:137–151
Gotoh T (1987a) Annual life cycle of the spider mite, *Eotetranychus uncatus* Garman (Acarina: Tetranychidae). Appl Entomol Zool 22:52–58
Gotoh T (1987b) Intraleaf distribution of *Panonychus ulmi* (Koch) (Acarina: Tetranychidae) on dwarf bamboo. Appl Entomol Zool 22:248–258
Gotoh T (1987c) Life-history parameters of two species of the genus *Eotetranychus* (Acarina: Tetranychidae) on deciduous trees. Appl Entomol Zool 22:45–51
Gotoh T (1987d) Comparison of life-history parameters of two forms of *Panonychus akitanus* Ehara (Acarina: Tetranychidae). Appl Entomol Zool 22:119–124
Gotoh T (1989) Seasonal life cycles and diapause in three species of *Schizotetranychus* (Acari: Tetranychidae). Appl Entomol Zool 24(4):407–417
Gotoh T, Shida T (2007) Life cycles and interactions in spider mites (Acari: Tetranychidae) on dwarf bamboo, *Sasa sinensis* (F. & S.) (Poaceae), in Japan. Int J Acarol 33:259–273
Gotoh T, Shinkaji T (1981) Critical photoperiod and geographical variation of diapause induction in the two-spotted spider mite, *Tetranychus urticae* Koch (Acarina: Tetranychidae), in Japan. Jpn J Appl Entomol Zool 25:113–118 (in Japanese with English summary)

Grafen A (1979) The hawk–dove game played between relatives. Anim Behav 27:905–907
Grafen A, Ridley M (1983) A model of mate guarding. J Theor Biol 102:549–567
Grandjean F (1948) Quelques caracteres des Tetranyques. Bull Mus Hist Nat Paris (2) 20:517–524
Gutierrez J, Helle W (1985) Evolutionary changes in Tetranychidae. In: Helle W, Sabelis MW (eds) Spider mites. Their biology, natural enemies and control, vol 1A. Elsevier, Amsterdam, pp 91–107
Hamilton WD (1964) The genetical evolution of social behaviour. I, II. J Theor Biol 7:1–52
Hamilton WD (1967) Extraordinary sex ratios. Science 156:477–488
Hamilton WD (1971) Geometry for the selfish herd. J Theor Biol 31:295–311
Hamilton WD (1972) Altruism and related phenomena, mainly in social insects. Annu Rev Ecol Syst 3:193–232
Hamilton WD (1979) Wingless and fighting males in fig wasps and other insects. In: Blum MS, Blum NA (eds) Sexual selection and reproductive competition in insects. Academic Press, London, pp 167–220
Hardy CWI, Mayhew PJ (19a98) Sex ratio, sexual dimorphism and mating structure in bethylid wasps. Behav Ecol Sociobiol 42:383–395
Harvey PH, Pagel M (1991) The comparative method in evolutionary biology. Oxford University Press, Oxford
Hazan A, Gerson U, Tahori AS (1973) Life history and life tables of the carmine spider mite. Acarologia 15:414–440
Hazan A, Gerson U, Tahori AS (1974) Spider mite webbing. I. The production of webbing under various environmental conditions. Acarologia 16:68–84
Helle W (1967) Fertilization in the two-spotted spider mite (*Tetranychus urticae*: Acari). Entomol Exp Appl 10:103–110
Helle W, Overmeer WPJ (1973) Variability in tetranychid mites. Annu Rev Entomol 18:97–120
Helle W, Pijnacker LP (1985) Parthenogenesis, chromosomes and sex ratio. In: Helle W, Sabelis MW (eds) Spider mites. Their biology, natural enemies and control, vol 1A. Elsevier, Amsterdam, pp 129–139
Helle W, Sabelis MW (eds) (1985) Spider mites. Their biology, natural enemies, and control, vols 1A, 1B. Elsevier, Amsterdam
Henter H (2003) Inbreeding depression and haplodiploidy: experimental measures in a parasitoid and comparisons across diploid and haplodiploid insect taxa. Evolution 57:1793–1803.
Herre EA (1985) Sex ratio adjustment in fig wasps. Science 228:896–898
Hines W, Maynard Smith J (1979) Games between relatives. J Theor Biol 79:19–30
Horita M, Chittenden AR, Sato Y, Saito Y (2004) Function of web box as an anti-predator barrier in the spider mite, *Schizotetranychus recki*. J Ethol 22:105–108
Hoy M (1985) Almonds (California). In: Helle W, Sabelis MW (eds) Spider mites. Their biology, natural enemies and control, vol 1B. Elsevier, Amsterdam, pp 299–310
Hoy MA, Groot JJR, van der Baan HE (1985) Influence of aerial dispersal on persistence and spread of pesticide-resistant *Metaseiulus occidentalis*. Entomol Exp Appl 37:117–131
Huffaker CB (1958) Experimental studies on predation: dispersion factors and predator–prey oscillations. Hilgardia 27:343–383
Huffaker CB, Shea KP, Herman SG (1963) Experimental studies on predation: complex dispersion and levels of food in an acarine predator–prey interaction. Hilgardia 34:305–329
Ims RA, Yoccoz NG (1997) Studying transfer processes in metapopulations: emigration, migration, and colonization. In: Hanski IA, Gilpin ME (eds) Metapopulation biology: ecology, genetics, and evolution. Academic Press, San Diego, pp 247–266
Ito K, Saito Y (2006) Effects of host-plant species on diapause induction of the Kanzawa spider mite (*Tetranychus kanzawai* Kishida). Entomol Exp Appl 121:177–184
Ito Y (1993) Behaviour and social evolution of wasps: the communal aggregation hypothesis. Oxford University Press, Oxford
Jackson DE, Hart AG (2009) Does sanitation facilitate sociality? Anim Behav 77:e1–e5 (doi:10.1016/j.anbehav.2008.09.013)
Jeppson LR, Keifer HH, Baker EW (1975) Mites injurious to economic plants. University of California Press, Berkeley.

Kasai A, Yano S, Takafuji A. (2005) Prey–predator mutualism in a tritrophic system on a camphor tree. Ecol Res 20:163–166

Kawakami Y, Gotoh SG, Ito K, Numata H (2008) Suppression of ovarian development and vitellogenin gene expression in the adult diapause of the two-spotted spider mite *Tetranychus urticae*. J Insect Physiol 55:70–77

Keller L (1995) Sterility by deleterious alleles and the evolution of sociality. J Theor Biol 174:467–469

Kojima K, Kobayshi D, Kobayashi S, Akitaya F, Narita H, Ishikawa N (1970) Report of pit-wall observations of snow cover in Sapporo, 1968–1969. Low Temp Sci Ser A 28:25–36

Koveos DS, Veerman A, Broufas GD, Exarhou A (1999) Altitudinal and latitudinal variation in diapause characteristic in the spider mite *Tetranychus urticae* Koch. Entomol Sci 24:607–613

Krantz GW, Walter DE (eds) (2009) A manual of acarology. Texas Technical University Press, Lubbock

Krebs JR, Davies NB (1987) An introduction to behavioural ecology, 2nd edn. Blackwell, Oxford

Kuno E (1987) Principles of predator-prey interaction in theoretical, experimental and natural population systems. Adv Ecol Res 16:249–337

Kuwahara Y (1990) Chemical studies on astigatid mites: opisthonotal gland secretions and body surface components with biological functions. J Pestic Sci 15:613–619

Kuwahara Y, My-Yen LT, Tominaga Y, Matsumoto K, Wada Y (1982) 1,3,5,7-Tetramethyldecyl formate, lardolure: aggregation pheromone of the acarid mite, *Lardoglyphus konoi* (Sasa et Asanuma) (Acarina: Acaridae). Agric Biol Chem 45:2283–2291

Leather SR, Walters KFA, Bale JS (1993) The ecology of insect overwintering. Cambridge University Press, London

Lees AD (1953) Environmental factors controlling the evocation and termination of diapause in the fruit tree red spider mite *Metatetranychus ulmi* Koch (Acarina: Tetranychidae). Ann Appl Biol 40:449–486

Lehmann L, Keller L (2006) The evolution of cooperation and altruism: a general frame work and a classification of models. J Evol Biol 19:1365–1376

Lin JZ, Nakao H, Saito Y (2002) A new species of *Tarsonemus* (Acari: Tarsonemidae), prey on eggs of *Schizotetranychus longus* Saito. J Acarol Soc Jpn 11:85–89

Lin JZ, Zhang ZQ (2002) Tarsonemidae of the world. Systematic & Applied Acarology Society, London

Lin N, Michener CD (1972) Evolution of sociality in insects. Q Rev Biol 47:131–159

Lynch M, Walsh B (1998) Genetics and analysis of quantitative traits. Sinauer, Sunderland, MA

Lindquist EE (1985) External anatomy. In: Helle W, Sabelis MW (eds) Spider mites. Their biology, natural enemies and control, vol 1A. Elsevier, Amsterdam, pp 3–28

Lotka AJ (1925) Elements of physical biology. Williams & Wilkins, BaltimoreMaeda T, Takabayashi J (2001) Patch-leaving decision of the predatory mite *Amblyseius womersleyi* (Acari: Phytoseiidae) based on multiple signals from both inside and outside a prey patch. J Insect Behav 14: 829–839

Maeda T, Takabayashi J (2001) Patch-leaving decision of the predatory mite *Amblyseius womersleyi* (Acari: Phytoseiidae) based on multiple signals from both inside and outside a prey patch. J Insect Behav 14:829–839

Maeda T, Takabayashi J (2005) Effects of foraging experiences on residence time of the predatory mite *Neoseiulus womersleyi* in a prey patch. J Insect Behav 18:323–333

Magalhañes S, Tudorache C, Montserrat M, van Maanen R, Sabelis MW, Janssen A (2004) Diet of intraguild predators affects antipredator behavior in intraguild prey. Behav Ecol 16:364–370

Matsuda H, Kawasaki K, Sgigesada, N, Teramoto E, Ricciardi LM (1985) Evolutionary and ecological stability of prey–predator systems with predator switching. In: Teramoto E, Yamaguti M (eds) Mathematical topics in population biology, morphogenesis and neurosciences. Springer-Verlag, Berlin, pp 172–181

Matsuura K, Fujimoto M, Goka K, Nishida T (2002) Cooperative colony foundation by termite female pairs: altruism for survivorship in incipient colonies. Anim Behav 64:167–173

Maynard Smith J (1964) Group selection and kin selection. Nature (Lond) 4924:1145–1147

Maynard Smith J (1976) Evolution and theory of games. Am Sci 64:41–45

Maynard Smith J (1982a) Evolution and the theory of games. Cambridge University Press, Cambridge
Maynard Smith J (1982b) The evolution of social behavior – a classification of models. In: King's College Sociobiology Group, Cambridge (ed) Current problems in sociobiology. Cambridge University Press, London, pp 29–44
Mayr E (1963) Animal species and evolution. Harvard University Press, Cambridge
McArthur RH, Wilson EO (1967) The theory of island biogeography. Princeton University Press, Princeton
McEnroe WD (1969) The photoresponse of the two-spotted spider mite, *Tetranychus urticae* Koch, to skyshine. Acarologia 11:281–283
McEnroe WD, Dronka K (1969) Eyes of the two-spotted spider mite. II. Behavioural analysis of the photoreceptors. Ann Entomol Soc Am 62:466–469
McGregor EA (1950) Mites of the family Tetranychidae. Am Midl Nat 44:257–420
McMurtry JA (1985) Avocado. In: Helle W, Sabelis MW (eds) Spider mites. Their biology, natural enemies and control, vol 1B. Elsevier, Amsterdam, pp 327–332
McMurtry JA, Croft BA (1997) Life-styles of phytoseiid mites and their roles in biological control. Annu Rev Entomol 42:291–321
McMurtry JA, Huffacker CB, Van de Vrie M (1970) Ecology of tetranychid mites and their natural enemies: their biological characters and the impact of spray practices. Hilgardia 40: 331–390
McMurtry JA, Johnson HG (1966) An ecological study of the spider mite *Oligonychus punicae* (Hirst) and its natural enemies. Hilgardia 37:363–402
Michod RE, Hamilton WD (1980) Coefficient of relatedness in sociobiology. Nature (Lond) 288:694–697
Mills LR (1973) Morphology of glands and ducts in the two-spotted spider mite, *Tetranychus urticae* Koch, 1836. Acarologia 15:218–236
Mitchell R (1973) Growth and population dynamics of a spider mite (*Tetranychus urticae* K., Acarina: Tetranychidae). Ecology 54:1349–1355
Mochizuki M, Takafuji A (1996) Effect of photoperiod and temperature on the diapause termination of the Kanzawa spider mite, *Tetranychus kanzawai* Kishida (Acari: Tetranychidae). J Acarol Soc Jpn 5:83–88
Montserrat M, Peña FDL, Hormaza JI, González-Fernández JJ (2008) How do *Neoseiulus californicus* (Acari: Phytoseiidae) females penetrate densely webbed spider mite nests? Exp Appl Acarol 44:101–106
Moraes DE, McMurtry JA, Denmark HA, Campos CB (2004) A revised catalog of the mite family Phytoseiidae. ZOOTAXA 434, Magnolia Press, Auckland
Mori H, Saito Y, Tho Y (1999a) Cooperative group predation in a sit-and-wait cheyletid mite. Exp Appl Acarol 23:643–651
Mori K (2000) Factors which caused variation in social systems of spider mites, *Schizotetranychus celarius* species group (Acari: Tetranychidae). PhD dissertation, Hokkaido University, Sapporo
Mori K, Saito Y (2004) Nest size variation reflecting anti-predator strategies in social spider mites of *Stigmaeopsis* (Acari: Tetranychidae). Behav Ecol Sociobiol 56:201–206
Mori K, Saito Y (2005) Variation in social behavior within a spider mite genus, *Stigmaeopsis* (Acari: Tetranychidae). Behav Ecol 16:232–238
Mori K, Saito Y (2006) Communal relationships in a social spider mite, *Stigmaeopsis longus* (Acari: Tetranychidae): an equal share of labor and reproduction between nest mates. Ethology 112:1–9
Mori K, Saito Y, Sakagami T (1999b) Effects of nest web and female attendance on survival of young in a subsocial spider mite, *Schizotetranychus longus* (Acari; Tetranychidae). Exp Appl Acarol 23:411–418
Mori K, Saito Y, Sakagami T, Sahara K (2005) Inbreeding depression of female fecundity by genetic factors retained in natural populations of a male haploid social mite (Acari: Tetranychidae) Exp Appl Acarol 36:15–23
Morimoto K, Furuichi H, Yano S, Osakabe M (2006) Web-mediated interspecific competition among spider mites. J Econ Entomol 99:678–684

Müller JK, Braunisch V, Hwang W, Eggert A-K (2006) Alternative tactics and individual reproductive success in natural associations of the burying beetle, *Nicrophorus vespilloides*. Behav Ecol 18:196–203

Nagelkerke CJ (1993) Evolution of sex allocation strategies of pseudo-arrhenotokous predatory mites (Acari: Phytoseiidae). PhD dissertation, University of Amsterdam, Amsterdam

Nagelkerke CJ, Sabelis MW (1996) Hierarchical levels of spatial structure and their consequences for the evolution of sex allocation in mites and other arthropods. Am Nat 148:16–39

Navajas M, Gutierrez J, Lagnel J (1996) Mitochondrial cytochrome oxidase I in tetranychid mites: a comparison between molecular phylogeny and changes of morphological and life history traits. Bull Entomol Res 86:407–417

Nickel JL (1960) Temperature and humidity relationships of *Tetranychus desertorum* Banks with special reference to distribution. Hilgardia 30:41–100

Nishiyama R, Wada Y, Mibu M, Yamaguchi Y, Shimogawara K, Sano H (2004) Role of a nonselective *de novo* DNA methyltransferase in maternal inheritance of chloroplast genes in the green alga, *Chlamydomonas reinhardtii*. Genetics 168:809–816

Norton RA, Kethley JB, Johnston DE, O'Connor BM (1993) Phylogenetic perspectives on genetic systems and reproductive modes of mites. In: Wrensch DL, Ebbert MA (eds) Evolution and diversity of sex ratio in insects and mites. Chapman & Hall, New York, pp 8–99

Okabe K, Makino S (2008) Parasitic mites as part-time bodyguards of a host wasp. Proc R Soc B 275:2293–2297

Oku K, Yano S, Osakabe M, Takafuji A (2005) Mating strategies of Kanzawa spider mite, *Tetranychus kanzawai* Kishida (Acari: Tetranychidae), in relation to the mating status of females. Ann Entomol Soc Am 98:625–628

Oku K, Yano S, Takafuji A (2003) Spider mite's use of a refuge during the quiescent stage in the presence of a predator. The Netherlands Entomological Society. Entomol Exp Appl 108:71–74

Oku K, Yano S, Takafuji A (2006) Host plant acceptance by the phytophagous mite *Tetranychus kanzawai* Kishida is affected by the availability of a refuge on the leaf surface. Ecol Res 21:446–452

Osakabe M, Hongo K, Funayama K, Osumi S (2006) Amensalism via webs causes unidirectional shifts of dominance in spider mite communities. Oecologia (Berl) 150:496–505

Osakabe M, Saito Y (1991) Morphological differences between the diapausing and non-diapausing forms of the citrus red mite, *Panonychus citri* (McGregor) (Acari: Tetranychidae). Bull Fruit Tree Res Stn 21:95–115

Osakabe M, Saito Y, Sakagami Y (1993) Protein differences detected by two-dimensional electrophoresis in the species complex of *Schizotetranychus celarius* (Banks). Exp Appl Acarol 17:757–764

Osakabe M, Sakagami Y (1994) RFLP analysis of ribosomal DNA in sibling species of spider mite, genus *Panonychus* (Acari: Tetranychidae). Insect Mol Biol 3:63–66

Overmeer WPJ (1985) Rearing and handling. In: Helle W, Sabelis MW (eds) Spider mites. Their biology, natural enemies and control, vol 1B. Elsevier, Amsterdam, pp 161–170

O'Neill SL, Hoffmann AA, Werren JH (1997) Influential passengers: inherited microorganisms and arthropod reproduction. Oxford University Press, New York

Pallini A, Janssen A, Sabelis MW (1997) Odour-mediated responses of phytophagous mites to conspecific and heterospecific competitors. Oecologia (Berl) 110:179–185

Penman DR, Cone WW (1972) Behaviour of male two-spotted spider mites in response to quiescent female deutonymphs and to web. Ann Entomol Soc Am 65:1289–1293

Potter DA, Wrensch DL, Johnston DE (1976) Guarding, aggressive behaviour and mating success in male two-spotted spider mites. Ann Entomol Soc Am 69:707–711

Pritchard AE, Baker EW (1955) A revision of the spider mite family Tetranychidae. Pac Coast Entomol Soc Mem Ser 2:1–472

Prokop Z, Stuglik M, Radwan J (2007) Male age, mating probability, and progeny fitness in the bulb mite. Behav Ecol 18:597–601

Queller DC (1985) Kinship, reciprocity and synergism in the evolution of social behaviour. Nature (Lond) 318:366–367

Queller DC (1994) Genetic relatedness in viscous populations. Evol Ecol 8:70–73

Radwan J (1993) The adaptive significance of male polymorphism in the acarid mite *Caloglyphus berlesei*. Behav Ecol Sociobiol 33:201–208

Radwan J (1995) Male morph determination in two species of acarid mites. Heredity 74:669–673

Radwan J (2007) Sexual selection and conflict in the bulb mite, *Rhizoglyphus robini* (Astigmata: Acaridae). Exp Appl Acarol 42:151–158

Radwan J, Unrug J, Tomkins JL (2002) Status-dependence and morphological trade-offs in the expression of a sexually selected character in the mite, *Sancassania berlesei*. J Evol Biol 15:744–752

Rayor LS, Taylor LA (2006) Social behavior in amblypygids, and a reassessment of arachnid social patterns. J Arachnol 34:399–421

Ridley M (1983) The explanation of organic diversity: The comparative method and adaptations for mating. Clarendon Press, Oxford.

Ridley M (1985) The problem of evolution. Oxford University Press, Oxford

Roeder C (1992) Sex ratio response of the two-spotted spider mite (*Tetranychus urticae* Koch) to changes in density under local mate competition. Can J Zool 70:1965–1967

Roeder C, Harmsen R, Mouldey S (1996) The effects of relatedness on progeny sex ratio in spider mites. J Evol Biol 9:143–151

Sabelis MW (1985) Reproductive strategies. In: Helle W, Sabelis MW (eds) Spider mites. Their biology, natural enemies and control, vol 1A. Elsevier, Amsterdam, pp 265–278

Sabelis MW, Bakker FM (1992) How predatory mites cope with the web of their tetranychid prey: a functional view on dorsal chaetotaxy in the Phytoseiidae. Exp Appl Acarol 16:203–255

Sabelis MW, Dicke M (1985) Long-range dispersal and searching behaviour. In: Helle W, Sabelis MW (eds) Spider mites. Their biology, natural enemies and control, vol 1B. Elsevier, Amsterdam, pp 141–160

Sabelis MW, Janssen A (1993) Evolution of life-history patterns in the Phytoseiidae. In: Houck MA (ed) Mites. Ecological and evolutionary analyses of life-history patterns. Chapman & Hall, New York, pp 70–98

Sabelis MW, van de Baan HE (1983) Location of distant spider mite colonies by phytoseiid predators: Demonstration of specific kairomones emitted by *Tetranychus urticae* and *Panonychus ulmi*. Entomol Exp Appl 33:303–314

Sabelis MW, Vermaat JE, Groeneveld A (1984) Arrestment responses of the predatory mite, *Phytoseiulus persimilis*, to steep odour gradients of a kairomone. Physiol Entomol 9:437–446

Saito Y (1977a) Study on the spinning behavior of the spider mite (Acarina: Tetrnychidae). I. Method for the quantitative evaluation of the mite webbing, and the relationship between webbing and walking. Jpn J Appl Entomol Zool 21:27–34 (in Japanese with English summary)

Saito Y (1977b) Study on the spinning behavior of the spider mite (Acarina: Tetrnychidae). II. The process of silk secretion and the webbing on the various stages of *Tetranychus urticae* Koch under different environmental conditions. Jpn J Appl Entomol Zool 21:150–157 (in Japanese with English summary)

Saito Y (1979a) Comparative studies on life histories of three species of spider mites (Acrina: Tetranychidae). Appl Entomol Zool 14:83–94

Saito Y (1979b) Study on spinning behavior of spider mites. III. Response of mites to webbing residues and their preferences for particular physical conditions of leaf surface (Acarina: Tetranychidae). Jpn J Appl Entomol Zool 23:82–91 (in Japanese with English summary)

Saito Y (1983) The concept of "life types" in Tetranychinae. An attempt to classify the spinning behaviour of Tetranychinae. Acarologia 24:377–391

Saito Y (1985) Life types of spider mites. In: Helle W, Sabelis MW (eds) Spider mites. Their biology, natural enemies and control, vol 1A. Elsevier, Amsterdam, pp 253–264

Saito Y (1986a) Biparental defence in a spider mite (Acari: Tetranychidae) infesting *Sasa* bamboo. Behav Ecol Sociobiol 18:377–386

Saito Y (1986b) Prey kills predator: counter-attack success of a spider mite against its specific phytoseiid predator. Exp Appl Acarol 2:47–62

Saito Y (1987) Extraordinary effects of fertilization status on the reproduction of an arrhenotokous and sub-social spider mite (Acari: Tetranychidae). Res Popul Ecol 29:57–71

Saito Y (1990a) Two new spider mite species of the *Schizotetranychus celarius* complex (Acari: Tetranychidae). Appl Entomol Zool 25:389–396

Saito Y (1990b) Life history and food habit of *Typhlodromus bambusae* Ehara, a specific predator of *Schizotetranychus celarius* (Banks) (Acari: Phytoseiidae, Tetranychidae). Exp Appl Acarol 10:45–51
Saito Y (1990c) "Harem" and "non-harem" type mating systems in two species of subsocial spider mites (Acari, Tetranychidae). Res Popul Ecol 32:1–16
Saito Y (1990d) Factors determining harem ownership in a sub-social spider mite (Acari, Tetranychidae). J Ethol 8:37–43
Saito Y (1990e). Natural enemies counterattacked – in acari predator–prey interactions. In: Kiritani K, Shiga M (eds) Ecology of natural enemies. Tokai University Press, Tokyo, pp 38–45 (in Japanese and Korean)
Saito Y (1994a) Do males of *Schizotetranychus miscanthi* (Acari, Tetranychidae) recognize kin in male competition? J Ethol 12:15–18
Saito Y (1994b) Is sterility by deleterious recessives an origin of inequalities in the evolution of eusociality? J Theor Biol 166:113–115
Saito Y (1995a) Clinal variation in male-to-male antagonism and weaponry in a subsocial mite. Evolution 49:413–417
Saito Y (1995b) Sociobiological aspects of spider mite life types. J Acarol Soc Jpn 4:55–67
Saito Y (1995c) "Altruistic gene" a priori involves two different deleterious gene effects. A reply to Keller. J Theor Biol 174:471–472
Saito Y (1997) Sociality and kin selection in Acari. In: Choe JC, Crespi BJ (eds) The evolution of social behaviour in insects and arachnids. Cambridge University Press, Cambridge, pp 443–457
Saito Y (2000) Do kin selection and intra-sexual selection operate in spider mites? Exp Appl Acarol 24:351–363
Saito Y, Chittenden AR, Mori K, Ito K, Yamauchi A (2008a) An overlooked side effect of nest scattering behavior to decrease predation risk (Acari; Tetranychidae, Stigmaeidae). Behav Ecol Sciobiol 63:33–42
Saito Y, Ito K, Sakagami T (2005) Imaginal induction of diapause in several 'adult-female diapausing' spider mites. Physiol Entomol 30:96–101
Saito Y, Mori H (1975) The effects of pollen as an alternative food for three species of phytoseiid mites (Acarina: Phytoseiidae). Mem Fac Agric Hokkaido Univ 9:236–246
Saito Y, Mori K (2005) Where does male-to-male "aggression" compromise "cooperation"? Popul Ecol 47:159–166
Saito Y, Mori K, Chittenden AR, Sato Y (2000a) Correspondence of male-to-male aggression to spatial distribution of individuals in field populations of a subsocial spider mite. J Ethol 18:79–83
Saito Y, Mori K, Sakagami T, Lin J (2004) Reinstatement of the genus *Stigmaeopsis* Banks, with descriptions of two new species (Acari, Tetranychidae). Ann Entomol Soc Am 97:635–646
Saito Y, Osakabe MH, Yasui Y (1993) A method for preparing permanent specimens of mites with Canada balsam. Appl Entomol Zool 28:593–597
Saito Y, Sahara K (1999) Two clinal trends in male-male aggressiveness in a subsocial spider mite (*Schizotetranychus miscanthi*). Behav Ecol Sociobiol 46:25–29
Saito Y, Sahara K, Mori K (2000b) Inbreeding depression by recessive deleterious genes affecting female fecundity of a haplo-diploid mite. J Evol Biol 13:668–678
Saito Y, Sakagami T, Sahara K (2002) Differences in diapause attributes between two forms distinguished by male-to-male aggression in a subsocial spider mite, *Schizotetranychus miscanthi* Saito. Ecol Res 17:645–653
Saito Y, Suzuki R (1987) Reexamination of several rearing methods for studying the life history of spider mites (Acari: Tetranychidae). Appl Entomol Zool 22:570–576
Saito Y, Takada T (2009) Conditions under which male-to-male aggression compromises cooperation in a game without learning. Open Evol J 3:1–8
Saito Y, Takahashi K (1980) Study on variation of *Schizotetranychus celarius* (Banks) I. Preliminary descriptions of morphological and life type variation. Jpn J Appl Entomol Zool 24:62–70 (in Japanese with English summary)
Saito Y, Takahashi K (1982) Study on variation of *Schizotetranychus celarius* (Banks). II. Comparison of mode of life between two sympatric forms (Acarina: Tetranychidae). Jpn J Ecol 32:69–78 (in Japanese with English summary)

Saito Y, Tsuji N, Chittenden AR (2008b) Sustainable pest management and moso bamboo forest conservation: a preliminary report. Proceedings of ICSA 2008, pp 309–316

Saito Y, Ueno J (1979) Life history studies on *Schizotetranychus celarius* (Banks) and *Aponychus corpuzae* Rimando as compared with other tetranychid mite species (Acarina: Tetranychidae). Appl Entomol Zool 14:445–452

Saito Y, Urano S, Nakao H, Amimoto K, Mori H (1996) A simulation model for predicting the efficiency of biological control of spider mites by phytoseiid predators. 1. The model. Jpn J Appl Entomol Zool 40:103–111 (in Japanese with English summary)

Sakagami T (2002) Phylogenetic analysis of subfamily Tetranychinae (Acari: Tetranychidae) in Japan, based on 28S ribosomal DNA sequences: does the Celarius species group belong to the genus *Schizotetranychus*? (in Japanese). PhD dissertation, Hokkaido Univ, Sapporo

Sakagami T, Saito Y, Kongchuensin M, Sahara K (2009) Molecular phylogeny of *Stigmaeopsis* with special reference to speciation through host-plant shift. Ann Entomol Soc Am 102:360–366

Santos MA, Laing JE (1985) Stigmaeid predators. In: Helle W, Sabelis MW (eds) Spider mites. Their biology, natural enemies and control, vol 1B. Elsevier, Amsterdam, pp 197–203

Sato Y, Saito Y (2006) Nest sanitation in social spider mites: interspecific differences in defecation behavior. Ethology 112:664–669

Sato Y, Saito Y (2007a) Can the extremely female-biased sex ratio of the social spider mites be explained by Hamilton's LMC? Ecol Entomol 32:597–602

Sato Y, Saito Y (2007b) Evolutionary view of waste management behavior using volatile chemical cues in social spider mites. J Ethol 26:267–272

Sato Y, Saito Y, Chittenden AR (2008) The parapatric distribution and contact zone of two forms showing different male-to-male aggressiveness in a social spider mite, *Stigmaeopsis miscanthi* (Acari: Tetranychidae). Exp Appl Acarol 44:265–276

Sato Y, Saito Y, Mori K (2000) Patterns of reproductive isolation between two groups of *Schizotetranychus miscanthi* Saito (Tetranychidae: Acari) showing different male aggression traits. Appl Entomol Zool 35:611–618

Sato Y, Saito Y, Sakagami T (2003) Rules for nest sanitation in a social spider mite, *Schizotetranychus miscanthi* Saito (Acari: Tetranychidae). Ethology 109:713–724

Savory T (1964) Arachinidae. Academic Press, London

Schausberger P (1999) Juvenile survival and development in *Euseius finlandicus*, *Typhlodromus pyri* and *Kampimodromus aberrans* (Acari: Phytoseiidae) feeding on con- and heterospecific immature life stages. Exp Appl Acarol 23:389–398

Schmidt G (1976) Der einfluss der von den beutetieren hinterlassenen spuren auf suchverhalten und suchherfolg von *Phytoseiulus persimilis* A.H. (Acarina: Phytoseiidae). Z Angew Entomol 82:216–218 (in German)

Schmidt JO (1990) Evolution of major defensive ensembles. In: Evans DL, Schmidt JO (eds) Insect defences. State University of New York Press, Albany, pp 1–2

Schmidt-Nielsen K (1984) Scaling: why is animal size so important? Cambridge University Press, Cambridge

Shimoda T, Kishimoto H, Takabayashi J, Amano H, Dicke M (2008) Comparison of thread-cutting behavior in three specialist predatory mites to cope with complex webs of *Tetranychus* spider mites. Exp Appl Acarol (doi 10.1007/s10493-008-9205-3)

Shinkaji N (1979) Geographical distribution of the citrus red mite, *Panonychus citri* and European red mite, *P. ulmi* in Japan. In: Rodriguez JG (ed) Recent advances in acarology, vol 1. Academic Press, New York

Smith RH, Shaw MR (1980) Haploid sex ratios and the mutation rate. Nature (Lond) 287:728–729

Southwood TRE (1977) Habitat, the template for ecological strategies? J Anim Ecol 46:337–365

Stouthamer R, Breeuwer JAJ, Hurst GDD (1999) *Wolbachia pipientis*: microbial manipulator of arthropod reproduction. Annu Rev Microbiol 53:71–102

Sugden R (1986) The economics of rights, co-operation and welfare. Blackwell, London

Suzuki S (1978) Index to Japanese Bambusaceae. Gakken, Tokyo, Japan

Takabayashi J, Dicke M (1996) Plant-carnivore mutualism through herbivore-induced carnivore attractants. Trends Plant Sci 1:109–113

Takafuji A, So PM, Tuno N (1991) Inter- and intra-population variations in diapause attribute of the two-spotted spider mite, *Tetranychus urticae* Koch, in Japan. Res Popul Ecol 33:331–344
Takahashi K (1987) Studies on morphological and ecological variations of *Schizotetranychus celarius* (Banks). PhD dissertation, Hokkaido University, Sapporo
Tallamy DW, Wood TK (1986) Convergence patterns in subsocial insects. Annu Rev Entomol 31:369–390
Tansky M (1978) Switching effect in prey-predator system. J Theor Biol 70:263–271
Taylor PD (1992) Altruism in viscous populations – an inclusive fitness model. Evol Ecol 6:352–356
Taylor PD, Bulmer MG (1980) Local mate competition and the sex ratio. J Theor Biol 86:409–419
Taylor RJ (1977) The value of clumping to prey. Oecologia (Berl) 30:285–294
Tilszer M, Antoszczyk K, Saiek N, Zajac E, Radwan J (2006) Evolution under relaxed sexual conflict in the bulb mite *Rhizoglyphus robini*. Evolution 60:1868–1873
Tomkins JL, Lebas NR, Unrug J, Radwan J (2004) Testing the status-dependent ESS model: population variation in fighter expression in the mite *Sancassania berlesei*. J Evol Biol 17:1377–1388
Treat AE (1958) Social organization in the moth ear mite (*Myrmonyssus phalaenodectes*). Proc 10th Int Cong Entomol 2:475–480
Treat AE (1975) Mites of moths and butterflies. Cornell University Press/Comstock, Ithaca, NY
Trivers R (1985) Social evolution. Benjamin & Cummings, Menlo Park, CA
Tsubaki Y, Siva Jothy MT, Ono T (1994) Re-copulation and post-copulatory mate guarding increase immediate female reproductive output in the dragonfly *Nannophya pygmeae* Rambur. Behav Ecol Sociobiol 35:219–225
Tsuji K (1990) Reproductive division of labor related to age in the Japanese queenless ant, *Pristomymex pungens*. Anim Behav 39:843–849
Uetz GW, Hieber CS (1997) Colonial web-building spiders: balancing the costs and benefits of group-living. In: Choe JC, Crespi BJ (eds) The evolution of social behavior in insects and arachnids. Cambridge University Press, Cambridge, pp 476–498
Usher MB, Davis PR (1983) The biology of *Hypoaspis aculeifer* (Canestrini) (Mesostigmata): is there a tendency towards social behaviour? Acarologia 24:243–250
Van Eydhoven GL, Helle W (1966) Sex abnormalities in the common spider mite (*Tetranychus urticae*) (Acari, Tetranychidae). Entomol Ber 26:204–208
Van Hoff JARAM (1990) Intergroup competition and conflict in animals and man. In: Van der Dennen J, Falger V (eds) Sociobiology and conflict. Evolutionary perspectives on competition, cooperation, violence and warfare. Chapman & Hall, London, pp 23–54
Van de Vrie M, Boersma A (1970) The influence of the predacious mite *Typhlodromus* (*A.*) *potentillae* (Garman) on the development of *Panonychus ulmi* (Koch) on apple grown under various nitrogen conditions. Entomophaga 15:291–304
Van de Vrie M, McMurtry JA, Huffaker CB (1972) Ecology of tetranychid mites and their natural enemies: a review. III. Biology, ecology, and pest status, and host-plant relations of tetranychids. Hilgardia 41:343–432
Van Wilgenburg E, Driessen G, Beukeboom LW (2006) Single locus complementary sex determination in Hymenoptera: an "unintelligent" design? Front Zool 3:1–15
Veerman A (1977) Photoperiodic termination of diapause in spider mites. Nature (Lond) 266:526–527
Veerman A (1985) Diapause. In: Helle W, Sabelis MW (eds) Spider mites. Their biology, natural enemies and control, vol 1A. Elsevier, Amsterdam, pp 279–316
Volterra V (1926) Variazioni e fluttuazioni del numero d'individui in specie animali conviventi. Mem R Accad Naz dei Lincei Ser VI 2:657–659
Vulinec K (1990) Collective security: aggregation by insects as a defense. In: Evans DL, Schmidt JO (eds) Insect defenses: adaptive mechanisms and strategies of prey and predators. State University of New York Press, Albany, pp 251–288
Waage JK (1979) Adaptive significance of postcopulatory guarding of mates and nonmates by male *Calopteryx maculata* (Odonata). Behav Ecol Sociobiol 6:147–154

Walter DE (1996) Living on leaves: mites, tomenta, and leaf domatia. Annu Rev Entomol 41:101–114

Walter DE, Proctor HC (1999) Mites: ecology, evolution and behaviour. CABI Publishing, Oxon

Wanibuchi K, Saito Y (1983) The process of population increase and patterns of resource utilization of two spider mites, *Oligonychus ununguis* (Jacobi) and *Panonychus citri* (McGregor), under experimental conditions (Acari: Tetranychidae). Res Popul Ecol 25:116–129

Weeks A R, Marec F, Breeuwer JAJ. (2001) A mite species that consists entirely of haploid females. Science 292:2479–2482

Werren JH, Colwell RK (1987) Labile sex ratios in wasps and bees. Bioscience 37:498–506

West SA, Murray MG, Machado CA, Griffin AS, Herre EA (2001) Testing Hamilton's rule with competition between relatives. Nature (Lond) 409:510–513

West SA, Pen I, Griffin AS (2002) Cooperation and competition between relatives. Science 296:72–75

Williams GC (1966) Adaptation and natural selection. A critique of some current evolutionary thought. Princeton University Press, Princeton

Wilson EO (1975) Sociobiology. Belknap Press of Harvard University Press, Cambridge

Wrensch DL (1993) Evolutionary flexibility through haploid males or how chance favors the prepared genome. In: Wrensch DL, Ebbert MA (eds) Evolution and diversity of sex ratio in insects and mites. Chapman & Hall, New York, pp 118–149

Wrensch DL, Kethley JB, Norton RA (1994) Cytogenetics of holokinetic chromosomes and inverted meiosis: keys to the evolutionary success of mites, with generalization on eukaryotes. In: Houck MA (ed) Mites. Chapman & Hall, New York, pp 282–343

Wright S (1922) Coefficient of inbreeding and relationship. Am Nat 56:330–338

Wynne-Edwards VC (1962) Animal dispersion in relation to social behaviour. Oliver & Boyd, Edinburgh & London

Xu L (1989) The biology of *Rhizomys pruinosus*. Acta Theriol Sin 4:100–105 (in Chinese: translated into English by W. Downs, Bilby Research Centre, Northern Arizona University, June 1989)

Yamamura N (2007) Conditions under which plants help herbivores and benefit from predators through apparent competition. Ecology 88:1593–1599

Yanagida H, Saito Y, Mori K, Chittenden AR (2001) Egg-depositing behavior as a predator avoidance tactic of *Yezonychus sapporoensis*. J Ethol 19:63–66

Yano S (2008) Collective and solitary behaviors of two-spotted spider mite (Acari: Tetranychidae) are induced by trail following. Ann Entomol Soc Am 101:247–252

Yano S, Takafuji A (2002) Variation in the life history pattern of *Tetranychus urticae* (Acari: Tetranychidae) after selection for dispersal. Exp Appl Acarol 27:1–10

Yasui Y (1988) Sperm competition of *Macrocheles muscaedomesticae* (Scopoli) (Acarina: Mesostigmata: Macrochelidae), with special reference to precopulatory mate guarding behavior. J Ethol 6:83–90

Yokoyama K (1932) New tetranychid mites attacking the mulberry leaf. Contribution I. Bionomics and external structures of *Tetranychus suginamensisi* n. sp. Bull Imp Sericult Exp Sta 8:229–287, 2 pls (in Japanese with English summary)

Young SS, Wrensch DL, Kongchuensin M (1986) Control of sex ratio by female spider mite. Entomol Exp Appl 40:53–60

Zhang YX (2002) Studies on factors causing outbreaks of pest mites in bamboo forests in Fujian, China. PhD dissertation, Hokkaido University, Sapporo

Zhang YX, Zhang QZ (eds) (2000) Biology and control of bamboo mites in Fujian. Special publication. Systematics & Applied Acarology Society, London

Zhang YX, Zhang ZQ, Saito Y, Liu QY, Ji J (2004) On the causes of mite pest outbreaks in mono- and poly-cultured moso bamboo forests. Chin J Appl Ecol 15:1161–1165 (in Chinese with English abstract)

Zhang YX, Zhang ZQ, Tong LX, Liu QY, Song MG (2000) Causes of pest mite outbreaks in bamboo forests in Fujian, China: analyses of mite damage in monoculture versus polyculture stands. Syst Appl Acarol Spec Publ 4:93–108

Zhang ZQ, BA Croft (1994) A comparative life history study of immature *Amblyseius fallacis*, *Amblyseius andersoni*, *Typhlodromus occidentalis*, and *Typhlodromus pyri* (Acari: Phytoseiidae) with a review of larval feeding patterns. Exp Appl Acarol 18:631–657

Index

S

T

MIX
Papier aus verantwortungsvollen Quellen
Paper from responsible sources
FSC® C105338

If you have any concerns about our products,
you can contact us on
ProductSafety@springernature.com

In case Publisher is established outside the EU,
the EU authorized representative is:
Springer Nature Customer Service Center GmbH
Europaplatz 3, 69115 Heidelberg, Germany

Printed by Libri Plureos GmbH
in Hamburg, Germany